T0214361

ASTROPHYSICAL FLOWS

Almost all conventional matter in the Universe is fluid, and fluid dynamics plays a crucial role in astrophysics. This new graduate textbook provides a basic understanding of the fluid dynamical processes relevant to astrophysics. The mathematics used to describe these processes is simplified to bring out the underlying physics. The authors cover many topics, including wave propagation, shocks, spherical flows, stellar oscillations and the instabilities caused by effects such as magnetic fields, thermal driving, gravity and shear flows. They also discuss the basic concepts of compressible fluid dynamics and magnetohydrodynamics.

The authors are Directors of the UK Astrophysical Fluids Facility (UKAFF) at the University of Leicester, and Editors of the Cambridge Astrophysics Series. This book has been developed from a course in astrophysical fluid dynamics taught at the University of Cambridge. It is suitable for graduate students in astrophysics, physics and applied mathematics, and requires only a basic familiarity with fluid dynamics.

JIM PRINGLE is Professor of Theoretical Astronomy and a Fellow of Emmanuel College at the University of Cambridge, and Senior Visitor at the Space Telescope Science Institute, Baltimore.

ANDREW KING is Professor of Astrophysics at the University of Leicester and a Royal Society Wolfson Research Merit Award holder. He is co-author of *Accretion Power in Astrophysics* (Cambridge University Press, third edition, 2002).

ASTROPHYSICAL FLOWS

J. E. PRINGLE
University of Cambridge

A. R. KING
University of Leicester

CAMBRIDGE
UNIVERSITY PRESS

CAMBRIDGE
UNIVERSITY PRESS

University Printing House, Cambridge CB2 8BS, United Kingdom

Published in the United States of America by Cambridge University Press, New York

Cambridge University Press is part of the University of Cambridge.

It furthers the University's mission by disseminating knowledge in the pursuit of education, learning and research at the highest international levels of excellence.

www.cambridge.org
Information on this title: www.cambridge.org/9781107693401

© J. Pringle and A. King 2007

First published 2007
First paperback edition 2014

A catalogue record for this publication is available from the British Library

ISBN 978-0-521-86936-2 Hardback
ISBN 978-1-107-69340-1 Paperback

Cambridge University Press has no responsibility for the persistence or accuracy of URLs for external or third-party internet websites referred to in this publication, and does not guarantee that any content on such websites is, or will remain, accurate or appropriate.

Contents

Preface

Almost all of the baryonic Universe is fluid, and the study of how these fluids move is central to astrophysics. This book originated in a 24-lecture course entitled 'Astrophysical Fluids' given by one of us (JEP) in Part III of the Mathematical Tripos at the University of Cambridge, comparable in level to a graduate course in the USA. The course was intended as a preparation for research, and the book reflects this. Preparing the lecture course and especially its booklist made it plain that there was a need to bring these ideas together in one place.

The book provides a brief coverage of basic concepts, but does assume some familiarity with undergraduate-level fluid dynamics, electromagnetic theory and thermodynamics. Our aim is to give a flavour of the fundamental fluid dynamical processes and concepts which an astrophysical theorist ought to know. To keep the book to a manageable size, we have had to be selective. In particular, we omit all discussion of dissipative fluid processes such as viscosity and magnetic diffusivity.

As well as covering a range of fluid dynamical concepts, we introduce some mathematical ideas and techniques. None of these is particularly deep or abstract, but some of the implementations do require some moderately heavy but straightforward algebra. Thus the reader will benefit from some familiarity with undergraduate-level mathematical methods, as well as some facility in mathematical manipulation. This takes practice and care, but more than anything it requires the ability to spot a mistake before proceeding too far.

Ideally, of course, one does not make mistakes, and some lecturers like to give their students the misleading impression that this is how research is done. In practice, errors occur all too frequently, and unfortunately some of these make their way into the research literature. The best method for finding errors is to understand the physical processes involved and how these processes are expressed in mathematical formulae. For this reason, this book emphasizes physical understanding and the extraction of relevant physical ideas from a mass of equations. To achieve this we often drastically simplify problems and keep only the physical processes of interest. For example, in the chapters on stellar oscillations we eliminate much of the heavy algebra which appears because real stars are spherical, and instead assume that stars are square (plane-parallel) or at worst (for rotating stars) cylindrical. This lets us get at the underlying physical processes without obscuring them with mathematics.

The problems at the ends of the chapters come both from the problem sheets associated with the course and from the examination questions set for it. They are intended to illustrate the course material further and also to introduce additional ideas. Thus they are an integral part of the book, and the determined reader will benefit from working through them.

1

The basic fluid equations

The subject of this book is how the matter of the visible Universe moves. Almost all of this matter is in gaseous form, and each gram contains of order 10^{24} particles (atoms, ions, protons, electrons, etc.), all moving independently except for interactions such as collisions. At first sight it might seem an impossible task to describe the evolution of such a complicated system. However, in many cases we can avoid most of this inherent complexity by approximating the matter as a *fluid*. A fluid is an idealized continuous medium with certain macroscopic properties such as density, pressure and velocity . This concept applies equally to gases and liquids, and we shall take the term fluid to refer to both in this book. The structure of matter at the atomic or molecular level is important only in fixing relations between macroscopic fluid properties such as density and pressure, and in specifying others such as viscosity and conductivity.

Describing a medium as a fluid is possible if we can define physical quantities such as density $\rho(\mathbf{r}, \mathbf{t})$ or velocity $\mathbf{u}(\mathbf{r}, \mathbf{t})$ at a particular place with position vector $\mathbf{r}$ at time t. For a meaningful definition of a 'fluid velocity' we must average over a large number of such particles. In other words, fluid dynamical quantities are well defined only on a scale l such that l is not only much greater than a typical interparticle distance, but also, more restrictively, much greater than a typical particle mean free path, λ_{mfp}.[†] Further, the concept of local fluid quantities is only useful if the scale l on which they are defined is much smaller than the typical macroscopic lengthscales L on which fluid properties vary. Thus to use the equations of fluid dynamics we require $L \gg l \gg \lambda_{\text{mfp}}$.

If this condition fails one should, strictly, not apply the fluid dynamical equations, but instead use concepts from plasma physics such as particle distribution functions. However, the huge additional complications and large physical uncertainties

[†] Roughly speaking, the mean free path is the average distance travelled by a typical particle before its trajectory is significantly deflected by another particle.

involved here mean that astrophysicists often apply fluid dynamical equations in situations where they are not strictly valid. The mean free path in astrophysical fluids is typically $\lambda_{\text{mfp}} \simeq 10^6 (T^2/n)$ cm, where T is the temperature (in K) and n is the number density (in cm^{-3}). In the centre of the Sun we have $T \simeq 10^7$ K, $n \simeq 10^{26}$ cm^{-3}, so $\lambda_{\text{mfp}} \sim 10^{-6}$ cm. This is far smaller than the solar radius $R_\odot = 7 \times 10^{10}$ cm, so the fluid approximation is very good. In the solar wind, however, we have $T \sim 10^5$ K, $n \sim 10$ cm^{-3} near the Earth's orbit, so that $\lambda_{\text{mfp}} \sim 10^{15}$ cm. This is far greater than the Sun–Earth distance, which is 1.5×10^{13} cm. Thus the fluid approximation does not apply well here, and the treatment of the interaction of the solar wind with the Earth's magnetosphere requires plasma physics. As a final example, the diffuse gas in a cluster of galaxies typically has $T \simeq 3 \times 10^7$ K, $n \simeq 10^{-3}$ cm^{-3}, and hence $\lambda_{\text{mfp}} \sim 10^{24}$ cm. This is of the same order as the physical size ~ 1 Mpc of a rich cluster. The fluid approximation is at best marginal for the diffuse regions of the cluster gas, but is nevertheless often used to gain a crude insight into its dynamics, heating and cooling. The dimensionless ratio λ_{mfp}/L of mean free path to typical flow lengthscale is called the Knudsen number Kn; $Kn \ll 1$ is a necessary condition for the validity of the fluid approximation. The results above show that $Kn \ll 1$ in the interior of the Sun, $Kn \gg 1$ in the solar wind, and $Kn \sim 1$ in cluster gas.

In this book we assume that the reader already has some familiarity with fluid dynamics, though not necessarily in an astrophysical context. For this reason the following derivation and discussion of the equations of fluid dynamics is brief. It is aimed mainly at establishing notation, as well as stressing those properties of fluids relevant to astrophysics which may be less familiar to fluid dynamicists from other fields.

1.1 Conservation of mass and momentum

The equations of fluid dynamics express conservation laws, and indeed one can use this basic property advantageously in devising numerical methods to solve them.

1.1.1 Mass conservation

Consider a fixed finite volume V within the fluid, bounded by the surface S. Then the mass of fluid contained within the volume is given by

$$\int_V \rho \, dV. \tag{1.1}$$

The mass contained in V can change only through a flux of fluid through the surface S. Thus conservation of mass implies the following:

$$\frac{d}{dt} \int_V \rho \, dV = -\int_S \rho \mathbf{u} \cdot d\mathbf{S}, \tag{1.2}$$

where $\mathbf{dS}$ is the (vector) element of area on the surface S. The volume is fixed, so we can take the derivative inside the term on the left-hand side (l.h.s.) and apply the divergence theorem to the term on the right-hand side (r.h.s.) to obtain

$$\int_V \left\{ \frac{\partial \rho}{\partial t} + \text{div}(\rho \mathbf{u}) \right\} dV = 0. \tag{1.3}$$

Since the volume V is arbitrary, we conclude that the integrand must itself vanish, that is

$$\frac{\partial \rho}{\partial t} + \text{div}(\rho \mathbf{u}) = 0, \tag{1.4}$$

and, equivalently, in suffix notation

$$\frac{\partial \rho}{\partial t} + \frac{\partial}{\partial x_j}(\rho u_j) = 0. \tag{1.5}$$

1.1.2 Momentum conservation

The momentum equation is obtained in exactly the same way by considering the rate of change of the total momentum in the volume V, given by

$$\frac{d}{dt} \int_V \rho \, \mathbf{u} \, dV. \tag{1.6}$$

The additional complication here is that as well as considering the flux of momentum across the surface S, we must take account of both the body force per unit volume f_i acting on the fluid and the surface stress given by an appropriate stress tensor T_{ij}. The momentum equation is then given by

$$\frac{\partial}{\partial t}(\rho u_i) + \frac{\partial}{\partial x_j}(\rho u_i u_j) = f_i + \frac{\partial}{\partial x_j}[T_{ij}]. \tag{1.7}$$

In this book we consider two main contributors to the body force. First we write the gravitational force as follows:

$$f_i = -\rho \frac{\partial \Phi}{\partial x_i}, \tag{1.8}$$

where the gravitational potential Φ is related to the density through Poisson's equation:

$$\nabla^2 \Phi = 4\pi G \rho, \tag{1.9}$$

where G is the gravitational constant. Second we take the magnetic force in the following form:

$$f_i = (\mathbf{j} \wedge \mathbf{B})_i, \tag{1.10}$$

where $\mathbf{j}$ is the current and $\mathbf{B}$ is the magnetic field.

We shall also briefly consider the electric force,

$$f_i = \rho_Q E_i, \tag{1.11}$$

where ρ_Q is the electric charge density and $\mathbf{E}$ is the electric field.

We define the stress tensor as follows. Consider an infinitesimal vector surface element $\mathbf{dS}$ within the fluid, where by convention the magnitude of the vector is the area of the surface element and the direction of the vector is normal to the surface element. Then the surface element is subject to a surface force $\mathbf{F}$ given by

$$F_i = T_{ij}\, dS_j. \tag{1.12}$$

We note that since both $\mathbf{dS}$ and $\mathbf{F}$ are vectors, then by the quotient rule T_{ij} is a second-order tensor.

In this book the main contributor to the stress tensor that we consider is the pressure p in the form

$$T_{ij} = -p\delta_{ij}, \tag{1.13}$$

where we make use of the Kronecker delta. In Section 1.5 we shall also write the magnetic force as a stress tensor as follows:

$$m_{ij} = B_i B_j - \frac{1}{2}\delta_{ij}B_k B_k. \tag{1.14}$$

Although we do not consider viscous effects in this book, we note here that the viscous stress terms come from relating the viscous contribution to the stress tensor to the second-order tensor $\partial u_i/\partial x_j$. This contains information about the relative flow of neighbouring fluid elements and is called the (rate of) strain tensor. Physically this expresses the fact that microscopic (especially thermal) motions within the ensemble of gas particles can transport momentum over distances of order the mean free path.

Finally, using the mass conservation equation, eq. (1.4), to replace the term $\partial\rho/\partial t$, we obtain the momentum equation (or the equation of motion of the fluid) in the following form:

$$\frac{\partial u_i}{\partial t} + u_j\frac{\partial u_i}{\partial x_j} = -\frac{1}{\rho}\frac{\partial p}{\partial x_i} - \frac{\partial \Phi}{\partial x_i} + \frac{\partial m_{ij}}{\partial x_j}. \tag{1.15}$$

1.2 The Lagrangian derivative

We can consider the evolution of a fluid quantity like the density $\rho(\mathbf{r}, t)$ in two ways. The partial derivative $\partial\rho/\partial t$ used above measures the way ρ changes with time t at a fixed position $\mathbf{r}$. But it is often more useful to consider the rate of change of the density of a particular fluid element as it moves with the fluid. This rate is called the Lagrangian derivative and is denoted by $D\rho/Dt$. We need to establish the relationship between these two concepts.

Suppose that a particular fluid element is at position $\mathbf{r}_0$ at time $t = 0$, and at a later time t is at a new position $\mathbf{r}(\mathbf{r}_0, t)$. Then the velocity of the fluid element is given by

$$\mathbf{u} = \frac{\partial}{\partial t} \mathbf{r}(\mathbf{r}_0, t), \tag{1.16}$$

where the partial derivative is taken at fixed $\mathbf{r}_0$. The Lagrangian derivative of (for example) the density of that particular fluid element is then simply given by

$$\frac{D\rho}{Dt} = \frac{\partial}{\partial t} \rho(\mathbf{r}(\mathbf{r}_0, t), t), \tag{1.17}$$

with the partial derivative taken at fixed $\mathbf{r}_0$. Since t appears in two places on the r.h.s. we may expect two terms in the derivative. Using the chain rule and the definition of $\mathbf{u}$ above we obtain

$$\frac{D\rho}{Dt} = \frac{\partial\rho}{\partial t} + \mathbf{u} \cdot \nabla\rho. \tag{1.18}$$

Thus, more generally the operator denoting the rate of change of a quantity following the fluid motion (the Lagrangian derivative) is given by

$$\frac{D}{Dt} = \frac{\partial}{\partial t} + \mathbf{u} \cdot \nabla. \tag{1.19}$$

1.3 Conservation of energy

We consider the heat content of a unit mass of fluid. In terms of thermodynamic quantities, a small change in the internal heat content of this unit mass is given by

$$T \, dS = de + p \, dV, \tag{1.20}$$

where T is the temperature, S is the entropy per unit mass, e is the internal energy per unit mass and V is the volume per unit mass. In terms of the density it is evident that $V = 1/\rho$, and thus

$$T dS = de - p\frac{d\rho}{\rho^2}. \tag{1.21}$$

Hence in a fluid flow, the rate of change of the heat content of a particular fluid element of unit mass is given by

$$T\frac{DS}{Dt} = \frac{De}{Dt} - \frac{p}{\rho^2}\frac{D\rho}{Dt}. \tag{1.22}$$

The heat content of a fluid element can change through effects of two types.

First, there may be heat flow into or out of the element. We shall refer to this generically as 'conduction'. However, in the astrophysical context heat can be conducted both by gas particles (typically electrons, since they move faster than the ions) as in standard thermal conduction and also by photons (known as radiative transfer). In both cases, the heat flux **h** in units of energy per unit area per unit time can often be written in the following form:

$$\mathbf{h} = -\lambda \nabla \mathbf{T}, \tag{1.23}$$

which implies physically that the heat flux occurs down the temperature gradient at a rate proportional to some 'thermal conductivity' λ. We expect λ to be a function of thermodynamic variables such as T and ρ. This form of the heat flux is appropriate provided that the particles carrying the heat have mean free paths much smaller than the typical lengthscale L over which macroscopic fluid quantities change. For electrons or molecules this is equivalent to the requirements of the fluid approximation, whereas for photons it requires in addition that the fluid should be opaque ('optically thick') so that there are very large numbers of interactions between photons and the fluid over lengthscales L.

Second, there may be internal generation of heat. This can result from dissipation of kinetic energy by viscosity or dissipation of magnetic energy through resistivity (or electrical conductivity). We do not consider these processes in this book. In the astrophysical context internal energy can be generated by nuclear processes (such as nuclear energy generation in stars) and by a change in ionization of the fluid. It can also be caused by heat exchange with particles which have a low collision cross section, for example heating by cosmic rays in the interstellar medium and radiative heating and/or cooling in an optically thin gas. We shall denote the generation of internal energy by ϵ in units of energy per unit volume per unit time.

To convert from the rate of change of a unit mass of fluid (given by eq. (1.22)) to the rate of change per unit volume, we multiply by the mass per unit volume, i.e. the density. Thus the heat equation becomes

$$\rho T \frac{DS}{Dt} = -\mathrm{div}\,\mathbf{h} + \epsilon. \tag{1.24}$$

1.4 The equation of state and useful approximations

To complete the set of equations obtained so far we need a relationship of the form $p = p(\rho, T)$, which is the equation of state for the fluid. In this book we shall assume the simplest form of the relationship, namely the equation of state of a perfect gas,

$$p = \frac{\mathcal{R}}{\mu}\rho T, \tag{1.25}$$

where $\mathcal{R}$ is the gas constant and μ is the mean particle mass, assumed to be constant. We also note that

$$\frac{\mathcal{R}}{\mu} = c_p - c_V, \tag{1.26}$$

where $c_p = T(\partial S/\partial T)_p$ is the specific heat at constant pressure and $c_V = T(\partial S/\partial T)_V$ is the specific heat at constant volume. Alternatively this may be written as follows:

$$p = (\gamma - 1)\rho e, \qquad (1.27)$$

where $\gamma = c_p/c_V$ is the ratio of specific heats, and we note for a perfect gas that

$$e = c_V T. \qquad (1.28)$$

To understand the physics of a particular fluid dynamical situation it is often not necessary to include the full thermodynamic complexity of the fluid. In these cases we can simplify and/or circumvent the heat equation.

1.4.1 Incompressible approximation

The major difference between astrophysical fluids and those encountered in many terrestrial situations (including those encountered in many courses on fluid dynamics) is that astrophysical ones are highly compressible. However, in situations where fluid motions are slow compared with the sound speed, density gradients are quickly smoothed out and it is a useful approximation to treat the fluid as if it were incompressible. In physical terms this means that any particular element of the fluid does not change its density, which implies that

$$\frac{D\rho}{Dt} = 0. \qquad (1.29)$$

It is important to realise that this does not imply that the fluid itself has constant density, so we may **not** write $\rho = $ constant, unless the original fluid state has uniform density.

1.4.2 Adiabatic flow

If the flow occurs fast enough that no fluid element has time to exchange heat with its surroundings, and if energy generation within the fluid is negligible, the heat equation simplifies to

$$\frac{DS}{Dt} = 0. \qquad (1.30)$$

In other words, each fluid element evolves at constant entropy – it remains on the same adiabat.

At constant entropy we note that

$$\frac{Dp}{Dt} = \left(\frac{\partial p}{\partial \rho}\right)_S \frac{D\rho}{Dt}, \qquad (1.31)$$

and that

$$\left(\frac{\partial p}{\partial \rho}\right)_S = \frac{c_p}{c_V} \left(\frac{\partial p}{\partial \rho}\right)_T. \tag{1.32}$$

Since for a perfect gas

$$\left(\frac{\partial p}{\partial \rho}\right)_T = \frac{p}{\rho}, \tag{1.33}$$

on using $\gamma = c_p/c_V$ we obtain

$$\frac{D}{Dt} \ln p = \gamma \frac{D}{Dt} \ln \rho. \tag{1.34}$$

Thus for adiabatic flow we may assume that

$$\frac{D}{Dt}(p/\rho^\gamma) = 0. \tag{1.35}$$

We note again that this does not imply that the entropy of the fluid is constant everywhere. But in this case if the fluid is initially isentropic (has uniform entropy) then it remains so.

1.4.3 Barotropic flow

We can avoid using the heat equation, and therefore simplify the analysis, by assuming that pressure is solely a function of density, i.e. $p = p(\rho)$. This is a useful approximation when the detailed thermal properties of the fluid are not directly relevant to the dynamics under consideration. Barotropic flow is more general than isentropic flow, and includes isothermal flow (for which $p \propto \rho$) as well as the polytropic approximation to the equation of state (relevant to fully degenerate matter),

$$p = A\rho^{1+1/n}, \tag{1.36}$$

where A and n are constants and n is called the polytropic index.

1.5 The MHD approximation

Astrophysical fluids are usually highly ionized (and so highly conducting) and permeated by magnetic fields. Understanding the interaction between the fluid and the magnetic fields it contains is therefore often important. The usual treatment of this interaction uses the magnetohydrodynamics (MHD) approximation. We stress that this is an approximation and that, in common with the fluid approximation, it is often tempting to use it in contexts where its validity is stretched.

We start by considering a fluid flow with a typical flow lengthscale L and typical flow timescale T. The usual MHD approximation depends on the assumption that the resulting typical flow velocity U is much less than the speed of light, i.e.

$U \sim L/T \ll c$. The approximation stems from the use of Ohm's law applied locally in the frame of the fluid. Thus we need to be able to transform between the fields $(\mathbf{E}, \mathbf{B})$ in the inertial frame and the fields $(\mathbf{E}', \mathbf{B}')$ in the frame of the fluid, which is moving with velocity $\mathbf{u}$. These are related by the usual Lorentz transformation:

$$\mathbf{E}' = (1 - \gamma)\left(\frac{\mathbf{u} \cdot \mathbf{E}}{u^2}\right)\mathbf{u} + \gamma(\mathbf{E} + \mathbf{u} \wedge \mathbf{B}), \qquad (1.37)$$

and

$$\mathbf{B}' = (1 - \gamma)\left(\frac{\mathbf{u} \cdot \mathbf{B}}{u^2}\right)\mathbf{u} + \gamma\left(\mathbf{B} - \frac{1}{c^2}\mathbf{u} \wedge \mathbf{E}\right), \qquad (1.38)$$

where

$$\gamma = \left(1 - \frac{u^2}{c^2}\right)^{-1/2}. \qquad (1.39)$$

Taking the low-velocity approximation $u^2 \ll c^2$ and neglecting terms of order (u^2/c^2), these relations become

$$\mathbf{E}' = \mathbf{E} + \mathbf{u} \wedge \mathbf{B} \qquad (1.40)$$

and

$$\mathbf{B}' = \mathbf{B}. \qquad (1.41)$$

The time evolution of the magnetic field is determined from the Maxwell equation,

$$\frac{\partial \mathbf{B}}{\partial t} = -\operatorname{curl} \mathbf{E}. \qquad (1.42)$$

By comparing dimensional quantities on each side of the equation we see that to order of magnitude $B/T \sim E/L$, or equivalently $E \sim (L/T)B \sim UB$.

The second relevant Maxwell equation is as follows:

$$\mu_0^{-1}\operatorname{curl} \mathbf{B} = \mathbf{j} + \epsilon_0\frac{\partial \mathbf{E}}{\partial t}. \qquad (1.43)$$

The second term on the r.h.s. is the displacement current, which permits the propagation of electromagnetic waves in vacuum with speed c, where $c^2 = 1/\epsilon_0\mu_0$. However, in the MHD approximation we neglect the displacement current. This is because the ratio between the displacement current and the term on the l.h.s. is given to order of magnitude as $(\epsilon_0 E/T)/(B/\mu_0 L) \sim (E/B)(U/c^2) \sim U^2/c^2 \ll 1$. Thus in the MHD approximation, electromagnetic waves are excluded and the current is given by

$$\mathbf{j} = \mu_0^{-1}\operatorname{curl} \mathbf{B}. \qquad (1.44)$$

Since $\mathbf{B}' = \mathbf{B}$, it follows that the current in the frame of the fluid is given by

$$\mathbf{j}' = \mathbf{j}. \qquad (1.45)$$

In the frame of the fluid Ohm's law becomes $\mathbf{j}' = \sigma \mathbf{E}'$, where σ is the conductivity. In this book we make the additional assumption that the conductivity is infinite, which then implies that $\mathbf{E}' = 0$, i.e. that

$$\mathbf{E} = -\mathbf{u} \wedge \mathbf{B}. \tag{1.46}$$

Substituting this into eq. (1.42) we obtain the induction equation,

$$\frac{\partial \mathbf{B}}{\partial t} = \text{curl}(\mathbf{u} \wedge \mathbf{B}), \tag{1.47}$$

which describes the time evolution of the magnetic field in the ideal MHD approximation.

We also need to consider the electromagnetic force acting on the fluid. The Lorentz force is given by

$$\mathbf{f} = \rho_Q \mathbf{E} + \mathbf{j} \wedge \mathbf{B}. \tag{1.48}$$

The charge density ρ_Q is related to the electric field $\mathbf{E}$ through the following Maxwell equation:

$$\text{div}\,\mathbf{E} = \rho_Q/\epsilon_0. \tag{1.49}$$

Thus the ratio between the electric and magnetic contributions to the Lorentz force on the fluid is (using eq. (1.44)) to order of magnitude $(\epsilon_0 E^2/L)/(B^2/L\mu_0) \sim U^2/c^2$. Further, the current $\rho_Q \mathbf{u}$ supplied by the moving charge density is also $\sim U^2/c^2$ times the current $\mathbf{j}$. Thus in the MHD approximation we can neglect both the electric charge and the electric field, and the electromagnetic force on the fluid is (using eq. (1.44)) simply given by

$$\mathbf{f} = \mu_0^{-1}(\text{curl}\,\mathbf{B} \wedge \mathbf{B}). \tag{1.50}$$

We can write this as

$$f_i = \frac{\partial m_{ik}}{\partial x_k}, \tag{1.51}$$

where

$$m_{ik} = \mu_0^{-1}\left(B_i B_k - \frac{1}{2}B^2 \delta_{ik}\right), \tag{1.52}$$

and we have used the final Maxwell equation,

$$\text{div}\,\mathbf{B} = 0. \tag{1.53}$$

1.5.1 Notation and units

We can now see that in the MHD approximation the electric field does not appear in any of the equations. The magnetic field appears only in the induction equation and in the Lorentz force. The induction equation is already dimensionally consistent and so does not change if different units are used for $\mathbf{B}$. In the Lorentz force the magnetic field only enters in the dimensional combination $[B^2/\mu_0]$ and the field is measured in tesla. In cgs units the magnetic field is measured in gauss and this combination should be replaced by $[B^2/4\pi]$. Throughout the rest of this book we shall simplify the analysis and omit the factor of μ_0^{-1} (or of $1/4\pi$).

1.6 Some basic implications

Here we note some basic results which will prove useful in later chapters and which help to provide a simple mental picture of some of the results we shall derive.

1.6.1 Bernoulli equation for a non-magnetic barotropic fluid

For a barotropic fluid we have $p = p(\rho)$ and we can define the quantity $h = \int \mathrm{d}p/\rho$. Then, in a gravitational potential Φ, the momentum equation becomes

$$\frac{\partial \mathbf{u}}{\partial t} + (\mathbf{u} \cdot \nabla)\mathbf{u} = -\nabla h - \nabla \Phi. \tag{1.54}$$

Using the vector identity

$$(\mathbf{u} \cdot \nabla)\mathbf{u} = \nabla\left(\frac{1}{2}u^2\right) - \mathrm{curl}\,\mathbf{u}, \tag{1.55}$$

we can rewrite this as

$$\frac{\partial \mathbf{u}}{\partial t} - \mathbf{u} \wedge \mathrm{curl}\,\mathbf{u} = -\nabla\left(\frac{1}{2}u^2 + h + \Phi\right). \tag{1.56}$$

If the flow is steady, then taking the scalar product with $\mathbf{u}$ implies

$$\mathbf{u} \cdot \nabla\left(\frac{1}{2}u^2 + h + \Phi\right) = 0, \tag{1.57}$$

and thus that the quantity $(\frac{1}{2}u^2 + h + \Phi)$ is constant on streamlines.

1.6.2 Advection of vortex lines

Consider a small line element $\mathbf{dl}(\mathbf{r}, t)$ in the fluid connecting two neighbouring fluid elements at positions $\mathbf{r}$ and $\mathbf{r} + \mathbf{dl}$. Then as the fluid elements move, so does the line element $\mathbf{dl}$. It is straightforward to show (see Problem 1.9.1) that the evolution of the line element is governed by the following equation:

$$\frac{\mathrm{D}}{\mathrm{D}t}\mathbf{dl} = (\mathbf{dl} \cdot \nabla)\mathbf{u}. \tag{1.58}$$

We define the vorticity at a point in the fluid as

$$\omega = \mathrm{curl}\,\mathbf{u}. \tag{1.59}$$

We can think of the vorticity as describing the local rotation rate of the fluid. We can obtain some insight into how the vorticity behaves by comparing the motion of vortex lines with the way that the line element connecting two fluid elements moves.

Taking the curl of eq. (1.56) yields

$$\frac{\partial \omega}{\partial t} = \text{curl}(\mathbf{u} \wedge \omega). \tag{1.60}$$

We now use the vector identity for any two vectors $\mathbf{a}$ and $\mathbf{b}$:

$$\text{curl}(\mathbf{a} \wedge \mathbf{b}) = (\mathbf{b} \cdot \nabla)\mathbf{a} - (\mathbf{a} \cdot \nabla)\mathbf{b} + \mathbf{a}\,\text{div}\,\mathbf{b} - \mathbf{b}\,\text{div}\,\mathbf{a}, \tag{1.61}$$

to obtain

$$\frac{\partial \omega}{\partial t} + (\mathbf{u} \cdot \nabla)\omega - (\omega \cdot \nabla)\mathbf{u} + \omega\,\text{div}\,\mathbf{u} = 0. \tag{1.62}$$

Here we have used the identity for any vector $\mathbf{a}$ that $\text{div}(\text{curl}\,\mathbf{a}) = 0$, so that $\text{div}\,\omega = 0$. The mass conservation equation (eq. (1.4)) in the form

$$\frac{D\rho}{Dt} + \rho\,\text{div}\,\mathbf{u} = 0 \tag{1.63}$$

lets us eliminate $\text{div}\,\mathbf{u}$ and hence obtain the time-evolution equation for the vorticity in the following form:

$$\frac{D}{Dt}\left(\frac{\omega}{\rho}\right) = \left[\left(\frac{\omega}{\rho}\right) \cdot \nabla\right]\mathbf{u}. \tag{1.64}$$

By comparing this equation with eq. (1.58) we see that the quantity ω/ρ, variously known as the vortensity or the potential vorticity, is advected with the fluid.

1.6.3 Advection of magnetic field lines

Equation (1.47) describing the time evolution of magnetic field $\mathbf{B}$ is exactly similar to eq. (1.60) describing the evolution of the vorticity ω. Thus the same analysis can be applied to $\mathbf{B}$, and we obtain

$$\frac{D}{Dt}\left(\frac{\mathbf{B}}{\rho}\right) = \left[\left(\frac{\mathbf{B}}{\rho}\right) \cdot \nabla\right]\mathbf{u}. \tag{1.65}$$

Thus we can also conclude that the quantity $\mathbf{B}/\rho$ is advected with the fluid. In other words, in the MHD approximation, and in the absence of dissipation, the magnetic field lines are carried along with the fluid flow.

1.7 Conservation of energy

Finally in this chapter we consider the equations describing the conservation of energy. The equations are derived directly from those given above, and so in physical terms they contain no new information. However, it is instructive to see the combined energy equation in conservative form. To do this we take the equation describing the evolution of the thermal energy density ρe of the fluid, and add to it terms describing the evolution of the kinetic energy density $\frac{1}{2}\rho u^2$ and the magnetic energy density $\frac{1}{2}B^2$.

1.7.1 Kinetic energy

The rate of change of kinetic energy density is given by

$$\frac{\partial}{\partial t}\left(\frac{1}{2}\rho u_i u_i\right) = \frac{1}{2}u_i u_i \frac{\partial\rho}{\partial t} + \rho u_i \frac{\partial u_i}{\partial t}. \tag{1.66}$$

On the r.h.s. we use the mass conservation equation, eq. (1.4), to replace $\partial\rho/\partial t$ and use the momentum equation, eq. (1.15), to replace $\partial u_i/\partial t$. Combining various terms we then obtain

$$\frac{\partial}{\partial t}\left(\frac{1}{2}\rho u^2\right) = -\frac{\partial}{\partial x_i}\left[\left(p + \frac{1}{2}\rho u^2\right)u_i\right] - \frac{p}{\rho}\frac{D\rho}{Dt} + u_i\frac{\partial m_{ij}}{\partial x_j} - \rho u_i\frac{\partial\Phi}{\partial x_i}. \tag{1.67}$$

1.7.2 Magnetic energy

The rate of change of magnetic energy density is given by

$$\frac{\partial}{\partial t}\left(\frac{1}{2}B^2\right) = \mathbf{B}\cdot\frac{\partial\mathbf{B}}{\partial t}. \tag{1.68}$$

Using the Maxwell equation

$$\frac{\partial\mathbf{B}}{\partial t} = -\text{curl}\,\mathbf{E} \tag{1.69}$$

and the vector identity

$$\text{div}(\mathbf{E}\wedge\mathbf{B}) = \mathbf{B}\cdot\text{curl}\,\mathbf{E} - \mathbf{E}\cdot\text{curl}\,\mathbf{B}, \tag{1.70}$$

this becomes

$$\frac{\partial}{\partial t}\left(\frac{1}{2}B^2\right) = -\text{div}(\mathbf{E}\wedge\mathbf{B}) - \mathbf{E}\cdot\text{curl}\,\mathbf{B}. \tag{1.71}$$

We use the ideal MHD approximation $\mathbf{E} + \mathbf{u}\wedge\mathbf{B} = 0$, the Maxwell equation relating $\mathbf{j}$ and curl $\mathbf{B}$, and the definition of the stess tensor m_{ij} to obtain the equation in the following form:

$$\frac{\partial}{\partial t}\left(\frac{1}{2}B^2\right) = -\text{div}(\mathbf{E}\wedge\mathbf{B}) - u_i\frac{\partial m_{ij}}{\partial x_j}. \tag{1.72}$$

1.7.3 The combined energy equation

We can now combine the equations governing the time evolution of kinetic and magnetic energy densities with the equation governing the time evolution of thermal energy as follows:

$$\frac{\partial}{\partial t}(\rho e) = -\text{div}(\rho e \mathbf{u} - \lambda \nabla T) + \frac{p}{\rho}\frac{D\rho}{Dt} + \epsilon \tag{1.73}$$

to obtain a total energy equation in the form

$$\frac{\partial g}{\partial t} = -\text{div}\,\mathbf{q} + r, \tag{1.74}$$

where

$$g = \rho\left(e + \frac{1}{2}u^2\right) + \frac{1}{2}B^2, \tag{1.75}$$

$$q_i = \left[\rho\left(e + \frac{1}{2}u^2\right) + p\right]u_i - \lambda\frac{\partial T}{\partial x_i} + (\mathbf{E} \wedge \mathbf{B})_i \tag{1.76}$$

and

$$r = \epsilon - \rho\mathbf{u} \cdot \nabla\Phi. \tag{1.77}$$

Here g represents the various energy densities – thermal, kinetic and magnetic. (Recall that in the MHD approximation the electric energy density is negligible.) The vector quantity $\mathbf{q}$ in eq. (1.74) represents energy fluxes. In square brackets in the first term of eq. (1.76), in addition to the thermal and kinetic energies, there is a term $p\mathbf{u}$, which represents the $p\,dV$ work being done in compressing the fluid. There is also the conducted heat flux and the flux of electromagnetic energy (the Poynting flux). Finally, the quantity r in eqs. (1.74) and (1.77) represents heat loss/gain by the fluid. The first term ϵ represents local energy generation, for example by nuclear burning, and the second term represents gravitational energy released by flow in the gravitational potential Φ, here assumed to be fixed in time.

In the rest of this book we will use these equations to study a large variety of astrophysical fluid phenomena. We shall try throughout to discuss the simplest possible examples, embodying the essential physics in each case.

1.8 Further reading

Further discussion of the derivation and validity of the equations of fluid dynamics are to be found in Batchelor (1967, Chap. 1) and Landau & Lifshitz (1959, Chap. I). A derivation of the equations of magnetohydrodynamics (MHD) is given in Roberts (1967, Chap. 1), who also provides a clear description of the thermodynamic relations made use of here. More details of these are to be found in Lifshitz &

Pitaevskii (1980, Chap. 2). A description of the relationship between MHD and plasma physics is given by Sturrock (1994, Chaps. 11, 12).

1.9 Problems

1.9.1 (a) At time t, neighbouring fluid particles A and B are at position vectors $\mathbf{r}$ and $\mathbf{r} + \mathbf{dl}$, respectively. At time $t + \delta t$, particle A is at $\mathbf{r} + \delta t \mathbf{u}(\mathbf{r})$, where $\mathbf{u}(\mathbf{r})$ is the velocity field of the fluid. Similarly, particle B is at $\mathbf{r} + \mathbf{dl} + \delta t \mathbf{u}(\mathbf{r} + \mathbf{dl})$. Use this to show that the time evolution of the line element $\mathbf{dl}$ which joins A and B is given by

$$\frac{D}{Dt}\mathbf{dl} = (\mathbf{dl} \cdot \nabla)\mathbf{u}. \tag{1.78}$$

Show that in a barotropic fluid the specific vorticity, i.e. ω/ρ, obeys the same equation.

This shows that vortex lines are carried bodily along in an inviscid, barotropic fluid.

(b) The circulation C around a closed curve Γ is defined as follows:

$$C = \oint_\Gamma \mathbf{u} \cdot \mathbf{dr}. \tag{1.79}$$

If the curve Γ moves with the fluid (assumed to be inviscid and barotropic), show that C is a constant.

This is known as 'Kelvin's circulation theorem'.

(c) For a conducting barotropic fluid with zero magnetic diffusivity, show that

$$\frac{D}{Dt}\left(\frac{\mathbf{B}}{\rho}\right) = \left(\frac{\mathbf{B}}{\rho}\right) \cdot \nabla\mathbf{u}. \tag{1.80}$$

This shows that magnetic field lines are carried bodily along in a perfectly conducting fluid.

1.9.2 A smooth circular cylinder of radius a and height h contains fluid of uniform density ρ, rotating uniformly with angular velocity Ω about the axis of symmetry. Compute the vorticity ω.

(a) Show that in cylindrical polar coordinates, (R, ϕ, z), the velocity field given by

$$\mathbf{u} = (0, R\Omega(t), z/h), \tag{1.81}$$

applied for an appropriate time, describes a stretching of the cylinder to a height of $2h$ while keeping the density and the rotation uniform.

For incompressible flow it is found that stretching vortex lines leads to an increase in their strength. Use the vorticity equation to show that this is not the case here.

(b) Show that the flow field given by

$$\mathbf{u} = (-R/a, R\Omega(t), 0), \tag{1.82}$$

applied for an appropriate time, describes decreasing the radius of the cylinder to
$a/2$ while keeping the density and the rotation uniform. Use the vorticity equation
to show that in this case the vorticity does change.

(c) Show that both these results can be deduced simply from consideration of
conservation of angular momentum. Use an appropriate combination of the two
flow fields to show that the increase in strength of vortex lines as they are stretched
in an incompressible fluid is also just a consequence of the conservation of angular
momentum.

1.9.3 A simple model for a filament in the solar atmosphere considers gas supported by
a magnetic structure. The configuration is steady and two-dimensional in the (x, z)-
plane, with constant gravity $\mathbf{g} = (0, 0, -g)$. The magnetic field is given by $\mathbf{B} = (B_x(x), 0, B_z(x))$, and is such that $B_z \to \pm B_0$ as $x \to \pm\infty$. Show that $B_x(x)$ is a
constant.

Assuming that the gas is isothermal, with sound speed c_s, and that the density $\rho(x)$
is a function of x alone, show that

$$B_z(x) = B_0 \tanh\left\{\frac{gB_0 x}{2c_s^2 B_x}\right\} \tag{1.83}$$

and find $\rho(x)$.

Sketch the magnetic field lines in the (x, z)-plane, and indicate where the density
is highest.

2

Compressible media

In Chapter 1 we emphasized that one of the major differences between astrophysical flows and the typical flows encountered in the terrestrial or laboratory context is that astrophysical fluids are compressible. This means that pressure information takes a finite time to propagate through the fluid. Because this time is often comparable to flow timescales, this gives compressible flows a fundamentally different character. In such flows the sound speed plays a role similar in some respects to that of the speed of light in the theory of relativity. In particular, sound travel times express physical causality. Pressure changes cannot propagate upstream in a supersonic flow. Subtle differences from the causal structure of relativity arise because, unlike the speed of light, the sound speed is variable and depends on the local properties of the fluid.

It is important to remember that *all* flows are compressible at some level. While the incompressible approximation is extremely useful in studying most terrestrial flows, intuition based on it is often a misleading guide in the astrophysical context. Moreover the elaborate mathematical apparatus assembled to study incompressible flows has limited applicability to astrophysical flows. For example, in incompressible fluids the pressure is formally disconnected from the other fluid variables, and appears only in the equation of motion, and only through its gradient (this is a mathematical expression of the assumption that it can adjust instantaneously at each point). Thus, taking the curl of the equation of motion eliminates the pressure from much of the analysis. This explains the prominent role played by the vorticity in the study of incompressible flows. By contrast, the vorticity and associated concepts such as velocity potentials and stream functions are of relatively little use in studying astrophysical flows.

In this chapter we consider various properties that are basic to an understanding of flows in compressible media.

2.1 Wave propagation in uniform media

We start by considering the simplest mechanisms transmitting information in compressible media. If the media are uniform, and not subject to external forces, then the simplest waves are pressure waves (acoustic or sound waves) and magnetic waves. We consider each in turn.

2.1.1 Small-amplitude sound waves

We consider a fluid at rest (velocity $\mathbf{u}_0 = 0$), with uniform density ρ_0 and uniform pressure p_0. Each fluid element is now perturbed by moving it a small distance $\boldsymbol{\xi}(\mathbf{r}, t)$. The density then becomes $\rho(\mathbf{r}, t)$. We are interested in the small change in density, known as the density perturbation. In general, we need to take care at this point and distinguish between Eulerian and Lagrangian perturbations. We mention this distinction briefly here, and return to it in more detail later on.

The Eulerian density perturbation is the change in density at a particular coordinate point $\mathbf{r}$. It is given by

$$\rho'(\mathbf{r}, t) = \rho(\mathbf{r}, t) - \rho_0(\mathbf{r}, t). \tag{2.1}$$

The Lagrangian density perturbation is the change in density for the particular fluid element which was at point $\mathbf{r}$ prior to the perturbation. It is given by

$$\delta\rho(\mathbf{r}, t) = \rho(\mathbf{r} + \boldsymbol{\xi}, t) - \rho_0(\mathbf{r}, t). \tag{2.2}$$

By using a Taylor expansion for $\rho(\mathbf{r} + \boldsymbol{\xi}, t)$, we note that to first order in $\boldsymbol{\xi}$,

$$\delta\rho = \rho' + \boldsymbol{\xi} \cdot \nabla\rho_0. \tag{2.3}$$

Thus for the stationary uniform medium we are currently considering the two ways of viewing the perturbation are the same since $\nabla\rho_0 = 0$, and thus for the time being we may take $\delta\rho = \rho'$.

For the moment, we consider Eulerian perturbations, and assume $p = p_0 + p'$ with $p' \ll p_0$, $\rho = \rho_0 + \rho'$ with $\rho' \ll \rho_0$ and $\mathbf{u} = \mathbf{0} + \mathbf{u}'$ with $\mathbf{u}'$ assumed small in some appropriate sense. We then substitute these into the equation of mass conservation:

$$\frac{\partial\rho}{\partial t} + \mathbf{u} \cdot \nabla\rho + \rho\operatorname{div}\mathbf{u} = 0, \tag{2.4}$$

and retain only small quantities to first-order. This yields

$$\frac{\partial\rho'}{\partial t} + \rho_0\operatorname{div}\mathbf{u}' = 0. \tag{2.5}$$

Similarly, we substitute into the momentum equation as follows:

$$\frac{\partial\mathbf{u}}{\partial t} + (\mathbf{u} \cdot \nabla)\mathbf{u} = -\frac{1}{\rho}\nabla p \tag{2.6}$$

and retain only terms linear in small quantities to obtain

$$\frac{\partial \mathbf{u}'}{\partial t} = -\frac{1}{\rho_0} \nabla p'. \tag{2.7}$$

To solve these equations we need to know the relationship between p' and ρ'. Small perturbations in pressure and density usually give rise to corresponding small perturbations in the temperature T. For acoustic waves it is often reasonable to assume that the perturbations occur sufficiently fast such that there is no time for temperature perturbations to be affected by thermal conduction of heat. If so, then the perturbations are 'adiabatic' in the sense that individual fluid elements remain on the same adiabat throughout the variations in pressure and density. This implies that the Lagrangian entropy perturbation for each fluid element is zero. For fluid on a particular adiabat we have seen that $p \propto \rho^\gamma$. Thus we conclude that

$$\frac{\delta p}{p_0} = \gamma \frac{\delta \rho}{\rho_0}. \tag{2.8}$$

However, since for a uniform fluid Lagrangian and Eulerian perturbations are the same, we conclude that, in this case,

$$p' = \left(\frac{\gamma p_0}{\rho_0}\right) \rho'. \tag{2.9}$$

We can now obtain an equation for the relative density perturbation (ρ'/ρ_0). Recalling that ρ_0 is constant, we take the time derivative of eq. (2.5), and use eqs. (2.7) and (2.9) to obtain

$$\frac{\partial^2}{\partial t^2} \left(\frac{\rho'}{\rho_0}\right) = c_s^2 \nabla^2 \left(\frac{\rho'}{\rho_0}\right), \tag{2.10}$$

where the quantity c_s is a constant and is defined as

$$c_s^2 = \gamma \frac{p_0}{\rho_0}. \tag{2.11}$$

Equation (2.10) is instantly recognizable as the linear wave equation for the quantity ρ'/ρ_0 with wave propagation speed equal to c_s.

The sound speed c_s is a fundamental quantity characterizing a compressible fluid. It fixes the maximum rate at which information about pressure, density, velocity and temperature changes can pass through the fluid and modify its behaviour. We note that it is a local quantity defined at each point of the fluid and can vary with position and time. In general we can write it as follows:

$$c_s = \left(\frac{\partial p}{\partial \rho}\right)^{1/2}, \tag{2.12}$$

where the derivative is evaluated using the energy equation and equation of state relating p and ρ. For example, if we consider isothermal rather than adiabatic perturbations (a good approximation in some cases), then $p_0 \propto \rho_0$ and we obtain

$$c_s^2 = \frac{p_0}{\rho_0} \tag{2.13}$$

rather than eq. (2.11). The perfect gas law, eq. (1.25), shows that in both cases $c_s \propto T^{1/2}$. Thus, in general, hotter gases have higher sound speeds.

2.1.2 Fourier transforms and the dispersion relation

An alternative, and often simpler, way of coming to the same conclusion is to use Fourier transforms. We note that the three eqs. (2.5), (2.7) and (2.9) for the linearized quantities ρ', $\mathbf{u}'$ and p' are linear differential equations with coefficients which are constant in both space and time. Thus, if we Fourier transform in both space and time, the equations for the transformed quantities will be algebraic. Thus, for example, we may consider

$$\tilde{p}'(\mathbf{k}, \omega) = \int_{-\infty}^{\infty} p'(\mathbf{r}, t) \exp[-i(\omega t + \mathbf{k} \cdot \mathbf{r})] dt \, d^3\mathbf{k}, \tag{2.14}$$

and similarly for $\tilde{\rho}'$ and $\tilde{\mathbf{u}}'$. Equivalently, and more simply, we may note that since the Fourier transform of $\partial p'/\partial t$ is $i\omega\tilde{p}'$, and the Fourier transform of $\nabla p'$ is $i\mathbf{k}\tilde{p}'$, we can achieve the same result by substituting

$$p'(\mathbf{r}, t) \longrightarrow \tilde{p}'(\mathbf{k}, \omega) \exp[i(\omega t + \mathbf{k} \cdot \mathbf{r})], \tag{2.15}$$

together with the corresponding quantities for $\mathbf{u}'$ and p'. For clarity we now drop the tildes and also take as read the factor $\exp[i(\omega t + \mathbf{k} \cdot \mathbf{r})]$ throughout. The equations become

$$i\omega\rho' + \rho_0 i\mathbf{k} \cdot \mathbf{u}' = 0, \tag{2.16}$$

$$i\omega\mathbf{u}' + \frac{i\mathbf{k}}{\rho_0}p' = 0 \tag{2.17}$$

and

$$p' = c_s^2\rho'. \tag{2.18}$$

Taking the scalar product of eq. (2.17) with $\mathbf{k}$, and using eqs. (2.16) and (2.18) to eliminate p' and ρ', we obtain the following equation:

$$(\mathbf{k} \cdot \mathbf{u}')[\omega^2 - k^2 c_s^2] = 0, \tag{2.19}$$

where $k = |\mathbf{k}|$. Then, provided that $\mathbf{k} \cdot \mathbf{u}' \neq 0$, that is provided that $\nabla \cdot \mathbf{u}' \neq 0$, i.e. the perturbations are compressible, we obtain the following relationship:

$$\omega^2 = k^2 c_s^2. \tag{2.20}$$

This relationship between the (angular) wave frequency ω and the wavenumber $\mathbf{k}$ is known as a dispersion relation. Note that the period of the wave is $P = 2\pi/\omega$, the wavelength of the wave is $\lambda = 2\pi/k$ and the wavefronts are perpendicular to the vector $\mathbf{k}$. The phase velocity of the waves is $(\omega/k)\hat{\mathbf{k}}$, and this is the velocity of the wavefronts. The group velocity $\mathbf{v}_g = \partial\omega/\partial\mathbf{k}$. This is the velocity at which the waves propagate information, i.e. the 'news' of pressure, density, velocity changes, etc.

For these simple acoustic waves we see that the phase velocity and the group velocity are the same and that both have magnitude equal to the sound speed c_s. Finally, from eq. (2.17) we see that $\mathbf{k} \wedge \mathbf{u} = 0$, i.e. curl $\mathbf{u} = 0$, so the acoustic waves represent irrotational perturbations. We conclude that these waves are longitudinal waves, with no transverse component.

The dispersion relation gives us essentially all the information about the properties of the waves. We have found this relation by replacing differential equations with algebraic ones. This is a much simpler procedure for obtaining a description of the nature of wave-like motions.

Large parts of this book will make extensive use of Fourier analysis and dispersion relations in this way. This is particularly true where we deal with small perturbations, as occurs in discussions of stellar oscillations and the stability of various flows.

2.1.3 Waves in a magnetic medium

We have seen that information travels through a compressible medium at the local sound speed. If the medium also has a magnetic field there are other ways of communicating physical information through it.

We consider the same unperturbed fluid as before, with uniform density ρ_0, uniform pressure p_0 and zero velocity $\mathbf{u}_0$, and add a uniform magnetic field $\mathbf{B}_0$. We consider small perturbations as before (i.e. $\rho = \rho_0 + \rho'$, $p = p_0 + p'$ and small velocity $\mathbf{u}$) and now have to add the perturbation to the magnetic field in the form $\mathbf{B} = \mathbf{B}_0 + \mathbf{b}(\mathbf{r}, t)$, where $|\mathbf{b}| \ll |\mathbf{B}_0|$. We then substitute these into the relevant equations, using the equilibrium conditions that $\nabla\rho_0 = 0$, $\nabla p_0 = 0$, $\mathbf{u}_0 = 0$ and curl $\mathbf{B}_0 = 0$. We also assume that the perturbations are adiabatic so that, as before, we may write $p' = c_s^2\rho'$, where $c_s^2 = \gamma p_0/\rho_0$ is uniform and constant. Then the mass conservation equation,

$$\frac{\partial\rho}{\partial t} + \mathrm{div}(\rho\mathbf{u}) = 0, \qquad (2.21)$$

becomes

$$\frac{\partial\rho'}{\partial t} + \rho_0\,\mathrm{div}\,\mathbf{u} = 0, \qquad (2.22)$$

and the momentum equation,

$$\rho\frac{\partial \mathbf{u}}{\partial t} + \rho(\mathbf{u}\cdot\nabla)\mathbf{u} = -\nabla p - \mathbf{B}\wedge(\nabla\wedge\mathbf{B}), \tag{2.23}$$

becomes

$$\rho_0\frac{\partial \mathbf{u}}{\partial t} = -c_s^2\nabla\rho' + \mathbf{B}_0\wedge(\nabla\wedge\mathbf{b}), \tag{2.24}$$

while the induction equation,

$$\frac{\partial \mathbf{B}}{\partial t} = \nabla\wedge(\mathbf{u}\wedge\mathbf{B}), \tag{2.25}$$

becomes

$$\frac{\partial \mathbf{b}}{\partial t} = \nabla\wedge(\mathbf{u}\wedge\mathbf{B}_0). \tag{2.26}$$

We note that eq. (2.26) implies that

$$\frac{\partial}{\partial t}(\operatorname{div}\mathbf{b}) = 0. \tag{2.27}$$

We differentiate eq. (2.24) with respect to time, and use eqs. (2.22) and (2.26) to eliminate $\partial\rho'/\partial t$ and $\partial\mathbf{b}/\partial t$ to obtain a linear equation for the velocity perturbation $\mathbf{u}$:

$$\rho_0\frac{\partial^2\mathbf{u}}{\partial t^2} = c_s^2\nabla\{\rho_0\operatorname{div}\mathbf{u}\} - \mathbf{B}_0\wedge\{\nabla\wedge[\nabla\wedge(\mathbf{u}\wedge\mathbf{B}_0)]\}. \tag{2.28}$$

We can simplify this a little by defining a vector quantity $\mathbf{V}_A$ with dimensions of velocity as follows:

$$\mathbf{V}_A = \frac{\mathbf{B}_0}{\sqrt{\rho_0}}, \tag{2.29}$$

which we shall call the vectorial Alfvén velocity . Then the equation becomes

$$\frac{\partial^2\mathbf{u}}{\partial t^2} - c_s^2\nabla(\operatorname{div}\mathbf{u}) + \mathbf{V}_A\wedge\{\nabla\wedge[\nabla\wedge(\mathbf{u}\wedge\mathbf{V}_A)]\} = 0. \tag{2.30}$$

Since each term contains either two time derivatives or two space derivatives, this is clearly a wave equation of some sort. If the magnetic field is zero (i.e. $\mathbf{B}_0 = \mathbf{V}_A = 0$) then the equation reduces to the equation for simple acoustic waves that we had before. But the term involving the magnetic field, with its four cross products, considerably complicates things.

In this case it is simpler to investigate the properties of these waves by using Fourier transforms, or equivalently by substituting $\mathbf{u}(\mathbf{r}, t) = \mathbf{u}(\mathbf{k}, \omega)\exp[i(\omega t + \mathbf{k}\cdot\mathbf{r})]$. We noted before that this is equivalent to making the transformations

$\partial/\partial t \to i\omega$ and $\nabla \to i\mathbf{k}$. With these substitutions, eq. (2.30) becomes

$$-\omega^2 \mathbf{u} + c_s^2 \mathbf{k}(\mathbf{k} \cdot \mathbf{u}) - \mathbf{V_A} \wedge \{\mathbf{k} \wedge [\mathbf{k} \wedge (\mathbf{u} \wedge \mathbf{V_A})]\} = 0. \tag{2.31}$$

To simplify this we first expand the vector triple product in the square brackets, namely

$$[\mathbf{k} \wedge (\mathbf{u} \wedge \mathbf{V_A})] = [(\mathbf{k} \cdot \mathbf{V_A})\mathbf{u} - (\mathbf{k} \cdot \mathbf{u})\mathbf{V_A}], \tag{2.32}$$

and then similarly expand the two resulting vector triple products $\mathbf{V_A} \wedge \{\mathbf{k} \wedge \mathbf{u}\}$ and $\mathbf{V_A} \wedge \{\mathbf{k} \wedge \mathbf{V_A}\}$. This gives the equation in the following form:

$$[\omega^2 - (\mathbf{k} \cdot \mathbf{V_A})^2]\mathbf{u} - (c_s^2 + V_A^2)(\mathbf{k} \cdot \mathbf{u})\mathbf{k} + (\mathbf{k} \cdot \mathbf{V_A})(\mathbf{u} \cdot \mathbf{V_A})\mathbf{k} + (\mathbf{k} \cdot \mathbf{V_A})(\mathbf{k} \cdot \mathbf{u})\mathbf{V_A} = 0. \tag{2.33}$$

This equation allows us to find the dispersion relation for these waves. As it is linear in $\mathbf{u}$, we can in principle write it in the following form:

$$A_{ij}u_j = 0, \tag{2.34}$$

where the coefficients of the matrix A are functions of ω and $\mathbf{k}$. The dispersion relation is then given by

$$\det A = 0, \tag{2.35}$$

which provides a functional relationship between ω and $\mathbf{k}$. However, while this dispersion relation contains all the information we require, it clearly does not do so in a particularly transparent form. The complicated nature of the equation means that we should not necessarily expect the wavevector $\mathbf{k}$ and the group velocity $\mathbf{v_g} = \partial\omega/\partial\mathbf{k}$ to be in the same direction. We now consider two special cases, for which in fact the wavevector and the group velocities are parallel, and which serve to illustrate the general properties of the waves. We leave a more complicated example to the Problems at the end of the chapter.

2.1.3.1 Wavefronts parallel to the magnetic field

Here we consider the case in which the wavevector $\mathbf{k}$ is perpendicular to the unperturbed magnetic field, i.e. $\mathbf{k} \cdot \mathbf{B_0} = 0$, which implies $\mathbf{k} \cdot \mathbf{V_A} = 0$. In this case, eq. (2.33) simplifies to

$$\omega^2 \mathbf{u} - (c_s^2 + V_A^2)(\mathbf{k} \cdot \mathbf{u})\mathbf{k} = 0. \tag{2.36}$$

Note that this equation implies that $\mathbf{u}$ is parallel to $\mathbf{k}$, i.e. that the waves are longitudinal. We take the scalar (dot) product of this equation with $\mathbf{k}$, and remove the factor $\mathbf{u} \cdot \mathbf{k} \neq 0$, to obtain the dispersion relation:

$$\omega^2 = (c_s^2 + V_A^2)k^2. \tag{2.37}$$

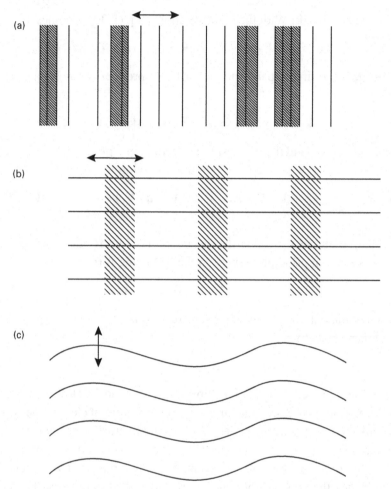

Fig. 2.1. Representative wave modes in fluids with magnetic fields. In each case the lines represent magnetic field lines, the shaded areas represent density enhancements, and the two-headed arrow represents the oscillatory motion of a fluid article. The waves are propagating horizontally across the page. (a) Fast magnetosonic waves propagating perpendicular to the field. Regions of high density and high field coincide. The field provides an extra contribution to the pressure. (b) Slow magnetosonic waves propagating along the field. Here the field is unperturbed and the waves are essentially just sound waves. (c) Alfvén waves propagating along the field. Here the density perturbations are zero and the perturbed field lines provide the restoring force.

These waves are exactly like the simple acoustic waves, except that the wave velocity is increased by the presence of the magnetic field (Fig. 2.1(a)). They are called fast magnetosonic waves and have a wave velocity of $v_{\text{fast}} = \sqrt{c_s^2 + B_0^2/\rho}$. Since $\mathbf{k} \cdot \mathbf{u} \neq 0$, the wave is compressive, and since $\mathbf{k} \parallel \mathbf{u}$ the wave velocity is perpendicular to the magnetic field. In the ideal MHD approximation the magnetic field is carried along with the fluid flow, so the effect of the wave motions is to try to change the distance between neighbouring magnetic field lines. The magnetic field resists this change, and the result is an additional restoring force, which acts exactly as an added magnetic pressure, and so provides an enhanced wave speed. We note that this wave exists even in the limit of small gas pressure (or temperature).

2.1.3.2 Wavefronts perpendicular to the magnetic field

If the wavefronts are orthogonal to the magnetic field we have $\mathbf{k} \parallel \mathbf{B}_0$, and we may set $\mathbf{k} = (k/V_A)\mathbf{V}_A$. Substituting this into eq. (2.33), we obtain

$$(k^2 V_A^2 - \omega^2)\mathbf{u} + \left(\frac{c_s^2}{V_A^2} - 1 \right) k^2 (\mathbf{u} \cdot \mathbf{V}_A)\mathbf{V}_A = 0. \tag{2.38}$$

We now have two possibilities. In the first case neither of the two coefficients vanishes, in which case we must have $\mathbf{u} \parallel \mathbf{V}_A$. Then we find after a little algebra that

$$\omega^2 = k^2 c_s^2. \tag{2.39}$$

These are the standard longitudinal acoustic waves (Fig. 2.1(b)). The wave motion is along the field lines, and is therefore unaffected by the presence of the field. In the presence of a magnetic field these correspond to slow magnetosonic waves.

In the second case both coefficients vanish. Then first we must have $\mathbf{u} \cdot \mathbf{V}_A = 0$, which since $\mathbf{k} \parallel \mathbf{V}_0$ implies that $\mathbf{k} \cdot \mathbf{u} = 0$. This means that div $\mathbf{u} = 0$, and therefore that the wave motion is incompressible. Second we have that

$$\omega^2 = k^2 V_A^2. \tag{2.40}$$

In this case the fluid motions displace the magnetic field sideways, producing a wave-like ripple in the field, in the same way as a violinist produces a sideways ripple in the string of the instrument being played (Fig. 2.1(c)). Just as a violin string tries to straighten itself, and produces a restoring force opposing the sideways shift, so the magnetic field tries to straighten itself, and so produces an analogous restoring force. These waves are transverse waves and are known as Alfvén waves. In this simple case the wave speed is simply the Alfvén speed $V_A = B_0/\sqrt{\rho_0}$. In an incompressible fluid these are the only waves present.

Thus, in general, in a magnetic medium there are three types of propagating disturbance: fast magnetosonic waves, slow magnetosonic waves and Alfén waves. This comes about because, in three dimensions, any initial perturbation of the medium can be described in terms of a set of three independent base vectors. In contrast to the case of a non-magnetic medium, in the presence of a field any initial disturbance perturbs either the field or the gas density (or both), and this leads to a propagating wave.

We have seen that magnetic wave phenomena introduce the fundamental propagation velocity – the Alfvén speed, V_A. The Alfvén speed plays a role in spreading magnetic perturbations similar to that of the sound speed for pressure waves. One might wonder what happens in the limit of very small density ρ_0, where V_A formally becomes infinite. In fact, once ρ_0 is small enough that V_A formally exceeds the speed of light c, it is clear that the treatment has become physically inconsistent. Then one has to revisit the approximations made in deriving the MHD equations.

2.2 Non-linear flow in one dimension

We have so far considered the properties of small-amplitude perturbations in compressible media. We have seen that such perturbations give rise to waves which propagate through the medium at some finite speed. This implies that information takes time to travel through the fluid. As we shall see, this finite timescale for the propagation of information can give rise to problems, for example if the fluid is moving at a velocity greater than the information propagation speed. To clarify such questions we consider the simplest case of one-dimensional compressible flow under pressure forces alone, with no restriction to small perturbations about equilibrium.

In one-dimensional flow the fluid quantities are functions of x and t only, and the flow velocity is in the x-direction with magnitude $u(x, t)$. We also assume that the flow is isentropic (the details of the flow are qualitatively similar for other choices of the relation between p and ρ). The isentropic assumption implies that throughout the flow $p = K\rho^\gamma$ for some constant K and $DS/Dt = 0$. We take the magnetic field to be zero.

With these assumptions, the mass conservation equation becomes

$$\frac{D\rho}{Dt} + \rho\frac{\partial u}{\partial x} = 0, \tag{2.41}$$

and the thermal equation (here conservation of entropy) becomes

$$\frac{D\rho}{Dt} = \frac{1}{c_s^2}\frac{Dp}{Dt}, \tag{2.42}$$

where the sound speed is given by $c_s^2 = \gamma p/\rho$. Eliminating $D\rho/Dt$ from these two equations, we obtain

$$\frac{1}{\rho c_s}\frac{\partial p}{\partial t} + \frac{u}{\rho c_s}\frac{\partial p}{\partial x} + c_s\frac{\partial u}{\partial x} = 0. \tag{2.43}$$

The momentum equation is given by

$$\frac{\partial u}{\partial t} + u\frac{\partial u}{\partial x} + \frac{1}{\rho}\frac{\partial p}{\partial x} = 0. \tag{2.44}$$

We now put these two equations together in an illuminating manner. First we add the two equations and gather terms to yield

$$\left[\frac{\partial u}{\partial t} + (u + c_s)\frac{\partial u}{\partial x}\right] + \frac{1}{\rho c_s}\left[\frac{\partial p}{\partial t} + (u + c_s)\frac{\partial p}{\partial x}\right] = 0. \tag{2.45}$$

Then we subtract the two equations and gather terms to yield

$$\left[\frac{\partial u}{\partial t} + (u - c_s)\frac{\partial u}{\partial x}\right] - \frac{1}{\rho c_s}\left[\frac{\partial p}{\partial t} + (u - c_s)\frac{\partial p}{\partial x}\right] = 0. \tag{2.46}$$

We note that these two equations are the same except for the change $c_s \leftrightarrow -c_s$. We can simplify these two equations still further by defining the quantities J_+ and J_-, known as Riemann invariants:

$$J_+ = u + \int \frac{dp}{\rho c_s} \tag{2.47}$$

and

$$J_- = u - \int \frac{dp}{\rho c_s}. \tag{2.48}$$

Then the two equations become

$$\left[\frac{\partial J_+}{\partial t} + (u + c_s)\frac{\partial J_+}{\partial x}\right] = 0 \tag{2.49}$$

and

$$\left[\frac{\partial J_-}{\partial t} + (u - c_s)\frac{\partial J_-}{\partial x}\right] = 0. \tag{2.50}$$

To interpret these equations we first need to consider the meanings of the quantities $[\cdots]$ in square brackets. To do this we consider the (x, t)-plane. A complete description of the flow is given by the functions $u(x, t)$ and $\rho(x, t)$ (or equivalently, since it is isentropic, $p(x, t)$) in this plane. Consider any function $f(x, t)$ (for example the density $\rho(x, t)$) in this plane, and consider a curve given by a monotonic function $x = \phi(t)$ in this plane. This curve describes the motion of something which moves at speed $d\phi/dt$ along the x-axis (see Fig. 2.2). Then the time derivative of $f(x, t)$ as seen by this something is given by

$$\left(\frac{df}{dt}\right)_\phi = \frac{\partial f}{\partial t} + \frac{d\phi}{dt}\frac{\partial f}{\partial x}. \tag{2.51}$$

We conclude therefore that on the curve C_+ in the (x, t)-plane defined by $x_+(t)$, where

$$\frac{dx_+}{dt} = u + c_s, \tag{2.52}$$

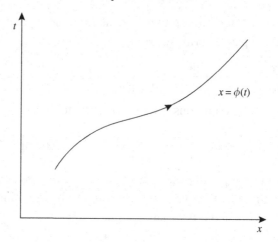

Fig. 2.2. Path of a point moving with speed $\mathrm{d}\phi/\mathrm{d}t$ along the x-axis.

eq. (2.49) implies that

$$\frac{\mathrm{d}J_+}{\mathrm{d}t} = 0 \tag{2.53}$$

and thus that $J_+ = $ constant. Similarly on the curve C_- in the (x, t)-plane defined by $x_-(t)$, where

$$\frac{\mathrm{d}x_-}{\mathrm{d}t} = u - c_\mathrm{s}, \tag{2.54}$$

eq. (2.50) implies that

$$\frac{\mathrm{d}J_-}{\mathrm{d}t} = 0 \tag{2.55}$$

and thus that $J_- = $ constant. The curves C_+ and C_- are known as characteristic curves, or simply characteristics.

We note further that using the fact that the flow is assumed to be isentropic, and thus that p, ρ and c_s are all mutually determined, we may write

$$J_+ = u + \frac{2c_\mathrm{s}}{\gamma - 1} \tag{2.56}$$

and

$$J_- = u - \frac{2c_\mathrm{s}}{\gamma - 1}. \tag{2.57}$$

Thus if we know u and c_s at any point, we can determine J_+ and J_-. Similarly, if we know J_+ and J_- at any point we can determine

$$u = \frac{1}{2}(J_+ + J_-) \tag{2.58}$$

and

$$c_s = \frac{\gamma - 1}{4}(J_+ - J_-). \qquad (2.59)$$

The analysis above reveals how information propagates through a compressible fluid in a remarkably simple way. Sound signals carry the information along the characteristics at the local sound speed c_s. In subsonic flow, information reaches any given point $x = x_0$ from smaller $x < x_0$ along the C_+ characteristic and from larger $x > x_0$ along the C_- characteristic.

We now discuss two particular sets of implications of this analysis.

2.2.1 Regions of influence

We consider a particular initial-value problem, where we suppose that at time $t = 0$ we have a complete knowledge of the fluid properties, i.e. we know $\rho(x, t = 0)$ (or, equivalently $p(x, t = 0)$ or $c_s(x, t = 0)$) and $u(x, t = 0)$. We then consider the solution $\rho(x, t)$ and $u(x, t)$ at later times $t > 0$. To be specific, consider the function $u(x, t)$ in the (x, t)-plane (we could also equally well consider the function $c_s(x, t)$). The initial condition at $t = 0$ is represented by the value of u along the x-axis $t = 0$. And, of course, the velocity structure at any particular later time $t = t_0$ is given by the value of u along the line $t = t_0$ parallel to the x-axis in the (x, t)-plane.

To form a physical idea of what the solution involves, we plot the characteristic curves C_+ and C_-, which can be thought of as starting on the x-axis $t = 0$ and propagating from there into the half-plane $t > 0$. We sketch these in Fig. 2.3. In the sketch we have assumed that the flow is subsonic, so that the C_+ curves propagate towards larger values of x, i.e. $dx_+/dt = u + c_s > 0$, and the C_- curves propagate towards smaller values of x, i.e. $dx_-/dt = u - c_s < 0$. Note that at

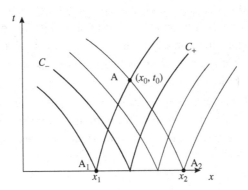

Fig. 2.3. Characteristic curves describing the one-dimensional motion of a compressible gas. The state of the gas at point A is determined by that at points $x_1 \le x \le x_2$ at time $t = 0$.

this stage this can only be a sketch, because in order to draw the curves accurately, we need to know the values of $u(x, t)$ and $c_s(x, t)$ at all points in the (x, t)-plane, or in other words we need to have already solved the problem! Now consider the value of $u(x_0, t_0)$ at the point A, shown in Fig. 2.3 at position $x = x_0, t = t_0$. As shown in Fig. 2.3, the characteristic curve C_+ which passes through the point (x_0, t_0) starts at the point A_1 with position $x_1 < x_0$ at time $t = 0$. Thus the Riemann invariant J_+ at A is determined by the initial $t = 0$ values of u and c_s at point A_1. That is,

$$J_+(x_0, t_0) = J_+(x_1, 0). \tag{2.60}$$

Similarly, the characteristic curve C_- which passes through the point (x_0, t_0) starts at the point A_2 with position $x_2 > x_0$ at time $t = 0$. Thus the Riemann invariant J_- at A is determined by the initial $t = 0$ values of u and c_s at A_2. That is,

$$J_-(x_0, t_0) = J_-(x_2, 0). \tag{2.61}$$

Then the values of u and c_s at point A are completely determined from these values of J_+ and J_- by using eqs. (2.58) and (2.59). Thus, for example,

$$u(x_0, t_0) = \frac{1}{2}[J_+(x_1, 0) + J_-(x_2, 0)], \tag{2.62}$$

with a similar equation to determine $c_s(x_0, t_0)$. We note further that, by a similar argument, the values of u and c_s at all points on the C_+ curve, and therefore the shape of the curve itself in the segment between A_1 and A, are determined by the initial values of u and c_s at points on the line segment $t = 0, x_1 \leq x \leq x_2$. Likewise, the values of u and c_s at all points on the C_- curve, and therefore the shape of the curve itself in the segment between A_2 and A, are determined by the initial values of u and c_s at points on the same line segment $t = 0, x_1 \leq x \leq x_2$. Thus the state of the gas at point A depends only on the state of the gas at points A_1 and A_2, together with the shapes of the characteristic curves through these points. But the shapes of the curves depend only on the initial $t = 0$ state of the gas at points $x_1 \leq x \leq x_2$. Thus the state of the gas at point A depends only on the initial state of the gas in a finite region. We can see that this comes about because information in a compressible gas travels only at a finite speed. Thus the state of the gas at point (x_0, t_0) can only depend on the state of those elements of the previous flow which have had time to communicate with it. From a physical point of view this is, in retrospect, obvious. This is a basic fact of compressible hydrodynamics, and indeed it is one which plays a large role in the development of numerical schemes for solving problems in compressible media.

We see that the system of characteristic curves makes explicit the physical causality implicit in the idea of information propagating at a finite speed, here that of sound. Very similar concepts appear in the theory of relativity, where the finite speed is of course that of light. Again there is a finite spatial region which is

able to influence any given event (point in space-time), and this is called the past
light cone of that point.

2.2.2 Development of shocks

We have seen above how to solve the evolution of a compressible fluid in one
dimension, at least in principle. Clearly similar considerations apply in more
dimensions. But it is also clear that the simple concept of characteristics which
move locally at speeds $\pm c_s$ with respect to the fluid runs into trouble if information
is forced to propagate through the fluid at a speed exceeding the local signal velocity.
This happens, for example, when an aircraft moves through air at supersonic speed.
The aircraft arrives at any point on its path before the sound waves it produces can
get there and tell the air to move out of its way. We know from experience that the
result is a shock wave, a region where the fluid quantities change on lengthscales
comparable with the mean free path. Here we examine first how such shock waves
arise, and then how to treat them in compressible fluid dynamics.

For a simple example of how shocks arise we consider a long tube of compressible
fluid (for example a gas) lying along the positive x-axis, with one end at $x = 0$.
We assume that the gas flow is one-dimensional, and that initially the fluid is at
rest, with $u = 0$ and $c_s = c_0 = $ constant. At time $t = 0$ we start to move a piston
into the fluid from the end at $x = 0$. We assume that the piston moves at constant
acceleration a, so that, as shown in Fig. 2.4 at time $t > 0$, the piston is at position
$x_p(t) = \frac{1}{2}at^2$ and has velocity $\dot{x}_p(t) = at$. We already see that, since we have set the

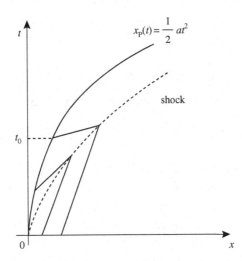

Fig. 2.4. Characteristic diagram for the one-dimensional motion of an accelerating
piston moving into a gas. A shock forms where the characteristics starting on the
piston intersect those starting in the stationary gas.

piston velocity to increase linearly with time and therefore eventually to exceed the speed of sound, we expect the method of solution outlined in the previous section to run into trouble.

Again we consider the evolution of the properties of the gas as given by the functions $u(x, t)$ and $c_s(x, t)$ in the (x, t)-plane. From Fig. 2.4 it is evident that the C_- characteristics, which start on the x-axis and move such that $dx_-/dt = u - c_s < 0$, fill the whole of the available space. Indeed, until they meet the effects of the advancing piston, they are straight lines with slope $dx/dt = -c_0$. On all these curves, the value of the Riemann invariant J_- is the same, and is equal to the value at time $t = 0$. We conclude that J_- is a constant throughout the fluid and is given by

$$J_- = -\frac{2c_0}{\gamma - 1}. \tag{2.63}$$

Thus, using the original expression (eq. (2.48)) for J_- we see that throughout the fluid c_s and u are related by the following expression:

$$c_s = c_0 + \frac{1}{2}(\gamma - 1)u. \tag{2.64}$$

Along the C_+ curves the quantity $J_+ = u + 2c_s/(\gamma - 1)$ is a constant. But since we have just shown that c_s depends only on u throughout the fluid, this implies that along each C_+ curve both $u = $ constant and $c_s = $ constant. Then since $dx_+/dt = u + c_s$ we see that the characteristic curves $x_+(t)$ are straight lines. A similar argument applies to the C_- curves.

Now consider the C_+ curve $x_+(t; t_0)$ which originates from the piston at time $t = t_0$, and therefore at position $x = \frac{1}{2}at_0^2$ and with velocity $u = at_0$. Using eq. (2.64), this implies that $c_s = c_0 + \frac{1}{2}(\gamma - 1)at_0$, and therefore that the characteristic curve is given by

$$\frac{d}{dt}x_+(t, t_0) = c_0 + \frac{1}{2}(\gamma + 1)at_0. \tag{2.65}$$

Using the initial conditions we can then integrate this to obtain

$$x_+(t, t_0) = \frac{1}{2}at_0^2 + \left\{c_0 + \frac{1}{2}(\gamma + 1)at_0\right\}(t - t_0). \tag{2.66}$$

We now note that each of the characteristic curves which originate on the piston is a straight line, and the quantities dx_+/dt *increase* with time t_0. Thus as shown in Fig. 2.4 in the (x, t)-plane the slopes of the lines *decrease* with time. This implies that they must intersect. This conclusion leads to a contradiction. We have already shown that at any point the fluid properties depend solely on the values J_+ and J_- of the Riemann invariants, which are constant along the characteristic curves C_+ and C_- passing through that point. But if there are two C_+ curves passing through a point, which is what must happen if two such curves intersect, then the fluid properties cannot be determined uniquely. But this does not make physical sense. It is clear that we can set up a physical experiment along the lines described here, and that the fluid evolution can be determined uniquely. What this implies is not that the fluid cannot make up its mind what to do, but rather that the mathematical method we are using to determine its evolution has broken down. As we mentioned at the start of this section, we know from experience what actually happens. The fluid properties change over a small region (the size of a few mean free paths) known as a shock. This invalidates the fluid approximation and the mathematical treatment given above in this small region.

2.2.3 Shock conditions

Just because the fluid approximation has broken down, this does not mean that we cannot determine what is going to happen. The shock itself is, by definition, confined to a region small compared with the lengthscale L on which fluid quantities would otherwise change. We can therefore idealize it as a mathematical surface, across which the fluid quantities change discontinuously. The basic fluid equations are conservation equations, so the changes in physical properties across such discontinuities must obey certain relationships.

2.2.3.1 Non-magnetic fluid

In general, the discontinuity is a curved surface, but if we consider a small enough portion of this surface we can treat it locally as if it is flat. It is simplest to work in the (instantaneous) frame in which the discontinuity is stationary. Thus we may assume that the discontinuity is in the plane $x = 0$ and that in this frame the flow is steady.

As shown in Fig. 2.5 we assume that the flow is in the positive x-direction. Thus in the upstream region, $x < 0$, the fluid has density ρ_1, pressure p_1 and velocity $\mathbf{u}_1$, with x-component $u_{1x} > 0$, and in the downstream region, $x > 0$, the fluid has density ρ_2, pressure p_2 and velocity $\mathbf{u}_2$, with x-component $u_{2x} > 0$. We write the unit normal to the plane of the shock, in the direction of the flow, as $\mathbf{n}$, where the components are given by $\mathbf{n} = (1, 0, 0)$.

Since the flow is steady, the mass conservation law becomes

$$\operatorname{div}(\rho \mathbf{u}) = 0. \tag{2.67}$$

By considering the divergence theorem applied to a short cylindrical volume (see Fig. 2.5(b)), in the limit that the length of the cylinder tends to zero, we find that the jump across the discontinuity is given by

$$[\rho \mathbf{u} \cdot \mathbf{n}]_1^2 = 0, \tag{2.68}$$

or, equivalently,

$$[\rho u_x]_1^2 = 0. \tag{2.69}$$

Compressible media

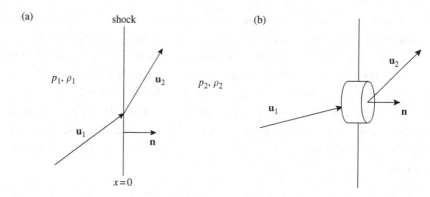

Fig. 2.5. (a) A stationary shock lies in the plane $x = 0$. Fluid with density ρ_1, pressure p_1 and velocity $\mathbf{u}_1$ flows into the shock from the half-space $x < 0$. The fluid flows away from the shock into the half-space $x > 0$ with density ρ_2, pressure p_2 and velocity $\mathbf{u}_2$. The unit vector $\mathbf{n}$ is normal to the shock in the direction of the flow. (b) The short cylindrical volume, as described in the text, to which we apply the divergence theorem in order to obtain the shock jump conditions.

In physical terms, this simply states that the mass flux going into the discontinuity must equal the mass flux coming out of it.

Similarly, for steady flow the momentum equation can be written as follows:

$$\frac{\partial}{\partial x_i}(\rho u_i u_j + p\delta_{ij}) = \cdots, \tag{2.70}$$

where $(\cdots)$ stands for the external force terms which are continuous across the jump. As before, we apply the divergence theorem (now in vector form) to the vanishingly small cylinder, to obtain the jump conditions:

$$[n_i(\rho u_i u_j + p\delta_{ij})]_1^2 = 0. \tag{2.71}$$

Of course, since momentum is a vector quantity, this is a vector equation. The three components of this equation can be written simply as follows:

$$[p + \rho u_x^2]_1^2 = 0, \tag{2.72}$$

$$[\rho u_x u_y]_1^2 = 0 \tag{2.73}$$

and

$$[\rho u_x u_z]_1^2 = 0. \tag{2.74}$$

We note immediately that eqs. (2.73) and (2.74) (ignoring the boring possibility $u_x = 0$), in combination with eq. (2.69), imply that

$$[u_y]_1^2 = 0 \tag{2.75}$$

and

$$[u_z]_1^2 = 0. \tag{2.76}$$

Thus the velocity components parallel to the shock are continuous. We may therefore, without loss of generality, assume that we can transform to a frame in which $u_y = u_z = 0$, and replace u_x by u.

In a similar manner we may apply the same procedure to the combined energy equation, eq. (1.74). We find that

$$[\mathbf{q} \cdot \mathbf{n}]_1^2 = 0, \tag{2.77}$$

where

$$\mathbf{q} = \left[\rho \left(e + \frac{1}{2} u^2 \right) + p \right] \mathbf{u}. \tag{2.78}$$

This implies

$$\left[\left\{ \rho \left(e + \frac{1}{2} u^2 \right) + p \right\} u_x \right]_1^2 = 0. \tag{2.79}$$

Again, making use of eq. (2.69), this implies

$$\left[\frac{1}{2} u^2 + e + \frac{p}{\rho} \right]_1^2 = 0. \tag{2.80}$$

The three eqs. (2.69), (2.72) and (2.80) are known as the Rankine–Hugoniot equations. The results of Problem 2.4.3 show that we can turn these into expressions for the velocity and density jumps (u_2/u_1), (ρ_2/ρ_2) in terms of the (upstream) *Mach number* $\mathcal{M} = u_1/c_s$ of the shock.

Physically what happens in a shock is that kinetic energy is turned into heat energy by dissipation. The result of Problem 2.4.4 shows this explicitly, and we find that the entropy of the fluid is higher on the downstream side of the shock. Also the sound speed is higher in this post-shock fluid, and comparable with the pre-shock fluid velocity . This is high enough to ensure that the decelerated post-shock fluid now moves *sub*sonically. By raising the sound speed in this way, the fluid is able to communicate the 'news' of an obstacle upstream into what would otherwise have been a supersonic flow. The high post-shock sound speed occurs because this part of the fluid is hotter. We see that the internal energy e increases across the shock.

Formally it is worth noting that the breakdown of the fluid approximation within the shock means that, for example, Bernoulli's theorem does not hold across it, precisely because the entropy and internal energy increase discontinuously at the expense of the fluid bulk motion (see eq. (2.80)). As we have seen, dissipation inevitably occurs in shocks, even though it may be negligible (as we have largely assumed) outside the shock, where the fluid approximation holds.

2.2.3.2 Magnetic fluid

We consider here the simple case in which the magnetic field in the upstream fluid is uniform. We use the arguments given above to justify considering the flow in the x-direction only. Then, if the magnetic field is perpendicular to the surface of discontinuity, i.e. $\mathbf{B} = (B, 0, 0)$, so that the flow is entirely along the magnetic field lines, then we have seen already that the flow is unaffected by the field. So, in this case, the jump conditions at the discontinuity are the same as in the non-magnetic case.

If the magnetic field is parallel to the surface of discontinuity, for example $\mathbf{B} = (0, B, 0)$, then we proceed as before, except that we need to include the magnetic terms in the momentum equation and the energy equation. The mass conservation equation is unchanged, and thus, as before, the jump condition is given by

$$[\rho u]_1^2 = 0. \tag{2.81}$$

Now the momentum equation becomes

$$\frac{\partial}{\partial x_i}(\rho u_i u_j + p\delta_{ij} + m_{ij}) = 0, \tag{2.82}$$

where

$$m_{ij} = B_i B_j - \frac{1}{2} B^2 \delta_{ij}. \tag{2.83}$$

Applying the vector divergence theorem as before, we find that the only non-zero component is the x-component, yielding

$$\left[\rho u^2 + p + \frac{1}{2} B^2 \right]_1^2 = 0. \tag{2.84}$$

The energy equation as before yields

$$[\mathbf{q} \cdot \mathbf{n}]_1^2 = 0, \tag{2.85}$$

where now, using the ideal MHD approximation $\mathbf{E} = -\mathbf{u} \wedge \mathbf{B}$,

$$\mathbf{q} = \left\{ \rho \left(e + \frac{1}{2} u^2 \right) + p \right\} \mathbf{u} + \mathbf{B} \wedge (\mathbf{u} \wedge \mathbf{B}). \tag{2.86}$$

Since in this case $\mathbf{u} \cdot \mathbf{B} = 0$ and $\mathbf{u} \cdot \mathbf{n} = u$,

$$\left[\left\{ \rho \left(e + \frac{1}{2} u^2 \right) + p + B^2 \right\} u \right]_1^2 = 0. \tag{2.87}$$

Finally we need to determine the jump in the size of B across the discontinuity. Since the field lines are carried along with the fluid, and since the field is perpendicular to the surface of discontinuity, it is straightforward to see what happens in this case. If, for example, the fluid is denser on the downstream side, then the fluid has been compressed there, and therefore the field lines have been pushed closer together. Imagine upstream in $x < 0$ two planes moving with the fluid which are parallel to the discontinuity and which are marked by particular field lines. Then when the planes have flowed through the discontinuity downstream in $x > 0$ the amount of fluid between the two planes is the same. Also the amount of magnetic flux between the two plane remains the same. This therefore implies that the quantity $(\mathbf{B}/\rho)$ remains unchanged. Thus the jump condition for the magnetic field is given by

$$\left[\frac{\mathbf{B}}{\rho}\right]_1^2 = 0. \tag{2.88}$$

2.2.4 Shock waves in general

We have seen that supersonic motion has a strong tendency to cause shock waves. It is not even necessary for some solid object (e.g. the piston moving into the cylinder that we discussed above) to move itself with supersonic speed for shocks to appear. Let us consider the same initially stationary isentropic gas-filled cylinder and simply move the piston a short distance into it, and then stop the piston entirely, before its own motion becomes supersonic with respect to the cylinder. Physically it is clear what happens. Sound waves travel ahead of the piston and tell the gas to move into the cylinder, raising its pressure and density. But these changes can only happen in the part of the gas close enough to the piston for sound to travel there in a given time. Further away from the piston the gas is unaware of its presence, and remains undisturbed.

One might imagine that once the piston stops moving the result would be a sound wave travelling smoothly into this gas. But from eq. (2.11) we can see that the sound speed in the gas varies with its density as $c_s \propto \rho^{(\gamma-1)/2}$. Since $\gamma > 1$, denser gas has a higher sound speed. Hence the compressed gas ahead of the piston has a higher sound speed than the undisturbed gas it is trying to push into. The result is that the compressed gas becomes still more compressed, further raising the sound speed within it, and further increasing the compression. The density contrast between the compressed and undisturbed gas increases until ρ and all other fluid quantities are changing significantly over a mean free path λ_{mfp}. This is exactly what we mean by a shock. Thus *any* motion of the piston into the gas ultimately leads to the formation of a shock at some distance ahead of it.

This process of density gradients steepening into shocks is already apparent in the characteristic diagram Fig. 2.4. Characteristics starting from the moving cylinder are 'refracted' to shallower slopes even *before* the cylinder itself begins to move supersonically. These refracted characteristics inevitably intersect the steeper ones in the undisturbed gas at larger x, showing, as expected, that a shock must eventually form.

Of course this does not mean that any isolated disturbance in a compressible medium eventually forms a shock wave. Human (and other) life would be very different if this were true! The special feature of the cylinder problem is that the gas is confined and can only move in one direction. In a three-dimensional case, the sound waves from an isolated disturbance spread roughly spherically, and this geometrical dilution reverses the tendency for the density to increase ahead of the sound waves. The sound wave spreads to large distances with decreasing amplitude before its energy is dissipated in microscopic motions in the gas, i.e. as heat.

Shock waves are extremely common in astrophysics. Gravitational and other forces can accelerate gas to supersonic speeds, or accelerate objects to move through gas supersonically. For example, hot stars produce winds which move highly supersonically into the interstellar medium, producing a system of shock waves, and supernova explosions cause even stronger effects. Similarly, the galaxies in a cluster move supersonically through the cluster gas, heating it via shocks. Any deceleration or deflection of supersonic gas inevitably leads to shocks, as when gas falls near radially down magnetic field lines on to a neutron star. The net effect of shocks is to turn bulk motion into heat, which in turn generally leads to radiation. Hence shocks are very often involved in producing much of the radiation we see from high-energy phenomena in the Universe.

2.3 Further reading

The relationship between Lagrangian and Eulerian perturbations is discussed further in Chapter 4. The concept of group velocity is described in more detail in Witham (1974, Chap. 11). The derivation of waves in magnetic media given here follows that given in Jackson (1998, Chap. 7); an alternative description is given by Sturrock (1994, Chap. 14). Non-linear flow of a compressible fluid in one dimension, the concept of characteristics and the development and treatment of shocks are discussed further in Zel'dovich & Raizer (1967, Chap. I) and in Landau & Lifshitz (1959, Chaps. IX, X). An analogy with traffic flow is described in Witham (1974, Chap. 3) and in Billingham & King (2000, Chap. 7).

2.4 Problems

2.4.1 Consider waves in a uniform compressible medium with uniform magnetic field **B** and sound speed c_s. In a Cartesian coordinate system, let the wavevector **k** be given by

$$\mathbf{k} = (0, 0, k), \tag{2.89}$$

the vectorial Alfvén velocity be given by

$$\mathbf{V_A} = V_A(0, \sin\theta, \cos\theta) \tag{2.90}$$

and the fluid velocity be given by

$$\mathbf{u} = (u_x, u_y, u_z)\exp\{i(\omega t - kz)\}, \tag{2.91}$$

where (u_x, u_y, u_z) is a constant vector.

Show that

$$u_x(\omega^2 - k^2 V_A^2 \cos^2\theta) = 0 \tag{2.92}$$

and find two similar equations for u_y and u_z.

If $u_x \neq 0$, show that the form of the motion is an incompressible (Alfvén) wave with phase velocity $V_A \cos\theta$.

If $u_x = 0$, show that

$$(\Omega^2 - \cos^2\theta)(\Omega^2 - \beta^2 - \sin^2\theta) - \sin^2\theta\cos^2\theta = 0, \tag{2.93}$$

where $\Omega = \omega/(kV_A)$ is the dimensionless phase velocity and $\beta = c_s/V_A$ is a dimensionless measure of the strength of the field.

Deduce that

$$\Omega^2(\beta, \theta) = \frac{1}{2}\{\beta^2 + 1 \pm [(\beta + 1)^2 - 4\beta^2\cos^2\theta]^{1/2}\}. \tag{2.94}$$

These represent the fast and slow magnetosonic waves.

For the case $\beta = 1$, plot the dimensionless phase velocities, $\Omega(\theta)$ as a function of θ. (See Parker (1979, Chap. 7).)

2.4.2 Along an infinite, straight, one-track road the local density of cars is $\rho(x, t)$ and the local velocity of cars (all assumed to be moving in the same direction) is $v(x, t)$. Discuss why it might be reasonable to assume that v is solely a function of ρ.

Making this assumption, show that

$$\frac{\partial\rho}{\partial t} + c(\rho)\frac{\partial\rho}{\partial x} = 0, \tag{2.95}$$

where the kinematic wave speed is defined by $c(\rho) = dQ/d\rho$ and $Q = \rho v$ is the local flux of cars.

Traffic flow along a particular highway can be fitted approximately for $\rho < \rho_{max}$ by

$$Q(\rho) = V_0\rho\log(\rho_{max}/\rho), \tag{2.96}$$

where $V_0 = 25$ kph and $\rho_{max} = 150$ vehicles km^{-1}.

Show that information propagates upstream at a speed V_0 relative to the local vehicle velocity .

Show that there is a maximum traffic flow which occurs at some density ρ_{crit}, corresponding to a critical speed v_{crit} of around 70 kph.

Describe the nature of the flow of traffic along this road. Show that if at some time the traffic density has $\rho = \rho_{crit}$ and $d\rho/dx > 0$ at some point, then there will in the future be stationary traffic at that point. Explain why the introduction of variable speed limits helps to ease traffic flow. (See Whitham (1974, Chap. 3) and Billingham & King (2000, Chap. 7).)

2.4.3 Use the Rankine–Hugoniot relations to show that

$$\frac{\rho_2}{\rho_1} = \frac{v_1}{v_2} = \frac{(\gamma + 1)\mathcal{M}_1^2}{(\gamma - 1)\mathcal{M}_1^2 + 2} \tag{2.97}$$

and

$$\frac{p_2}{p_1} = \frac{2\gamma \mathcal{M}_1^2}{\gamma + 1} - \frac{\gamma - 1}{\gamma + 1}, \tag{2.98}$$

where subscript 1 refers to upstream and subscript 2 to downstream of the shock, and $\mathcal{M}_1 = v_1/c_1$ is the Mach number of the shock.

Show also that $v_1 v_2 = c_*^2$, where the critical velocity c_* is defined by the following equation:

$$\frac{\gamma p_1}{(\gamma - 1)\rho_1} + \frac{1}{2}v_1^2 = \frac{\gamma + 1}{2(\gamma - 1)}c_*^2. \tag{2.99}$$

2.4.4 Use the Rankine–Hugoniot relations to show that the downstream Mach number $\mathcal{M}_2$ at a shock obeys the following:

$$\mathcal{M}_2^2 = \frac{(\gamma - 1)\mathcal{M}_1^2 + 2}{2\gamma \mathcal{M}_1^2 - (\gamma - 1)} \tag{2.100}$$

and show that this relation can be written as

$$X_1 X_2 = 1, \tag{2.101}$$

where

$$X = \frac{2\gamma}{\gamma + 1}(\mathcal{M}^2 - 1) + 1. \tag{2.102}$$

Express ρ_2/ρ_1 and p_2/p_1 in terms of X_1. What is the allowable range of X_1?

Show that the entropy change through the shock is given by

$$\frac{1}{c_V}(S_2 - S_1) = \ln X_1 - \gamma \ln\left[\frac{(\gamma + 1)X_1 + (\gamma - 1)}{(\gamma - 1)X_1 + (\gamma + 1)}\right] \tag{2.103}$$

and deduce that only compressive shocks ($\rho_2 > \rho_1$) occur in nature.

2.4.5 Show that the magnetohydrodynamics equations for a non-viscous, infinitely conducting fluid obeying the perfect gas law may be written as follows:

$$\frac{\partial \rho}{\partial t} + \operatorname{div}(\rho \mathbf{u}) = 0, \tag{2.104}$$

$$\frac{\partial}{\partial t}(\rho \mathbf{u}) + \operatorname{div}(\rho \mathbf{u}\mathbf{u} + pI - T) = -\rho \nabla \Phi \tag{2.105}$$

and

$$\frac{\partial}{\partial t}\left(\frac{1}{2}\rho u^2 + \rho e + \rho \Phi + \frac{1}{2}B^2\right)$$

$$+ \operatorname{div}\left[\rho \mathbf{u}\left(E + \frac{p}{\rho} + \frac{1}{2}u^2 + \Phi\right) + (\mathbf{B} \wedge \mathbf{u}) \wedge \mathbf{B})\right] = \rho \frac{\partial \Phi}{\partial t}, \tag{2.106}$$

where I is the unit tensor, and the tensor T is given by

$$T = \left(\mathbf{BB} - \frac{1}{2}B^2 I\right). \tag{2.107}$$

Show that in a steady hydromagnetic shock, in which the magnetic field and the flow velocity are normal to the shock front, the jump conditions across the shock front are the usual Rankine–Hugoniot conditions. Give a brief physical explanation of why the magnetic field plays no role in this case.

Now consider the jump conditions for a steady hydromagnetic shock in which the magnetic field is parallel to the shock front and the flow velocity is normal to it. Show that $B_1/\rho_1 = B_2/\rho_2$, where the subscripts 1 and 2 refer to pre- and post-shock velocities, respectively.

Show that the jump conditions concerning conservation of mass and momentum imply that

$$p_2 = p_1 + \rho_1 u_1^2\left(1 - \frac{\rho_1}{\rho_2}\right) + \frac{B_1^2}{2}\left(1 - \frac{\rho_2^2}{\rho_1^2}\right). \tag{2.108}$$

Use the jump condition concerning energy conservation to obtain another expression for p_2 in terms of the pre-shock variables and the ratio ρ_1/ρ_2.

Assuming that $\rho_1 \neq \rho_2$, deduce that $x = \rho_2/\rho_1$ is given by the following quadratic equation:

$$(2 - \gamma)V_{A1}^2 x^2 + [(\gamma - 1)u_1^2 + 2c_1^2 + \gamma V_{A1}^2]x - (\gamma + 1)u_1^2 = 0, \tag{2.109}$$

where $c_1^2 = \gamma p_1/\rho_1$ and $V_{A1}^2 = B_1^2/\rho$.

For a shock to exist we require that one of the roots is greater than unity. Show that this implies the following:

$$u_1^2 > c_1^2 + V_{A1}^2. \tag{2.110}$$

Give a physical interpretation of this condition. (See Field *et al.* (1968).)

2.4.6 A plane shock wave lies (in the frame of the shock) in the plane $x = 0$. The flow velocity is in the x-direction and is of magnitude U_L (U_R) to the left (right) of the shock, where left (right) corresponds to the half-space $x < 0$ ($x > 0$). In the same notation the densities are ρ_L (ρ_R), the pressures are p_L (p_R) and the energy densities are e_L (e_R). Assuming that the perfect gas law $p = (\gamma - 1)\rho e$ applies on each side of the shock, use the Rankine–Hugoniot relations to show that

$$\frac{\rho_R}{\rho_L} = \frac{(\gamma + 1)\mathcal{M}_L^2}{(\gamma - 1)\mathcal{M}_L^2 + 2},$$ (2.111)

where $\mathcal{M}_L$ is the Mach number in $x < 0$.

Deduce that $\rho_R > \rho_L \Leftrightarrow \mathcal{M}_L^2 > 1$, and hence that the flow must be supersonic on one side of the shock and subsonic on the other.

Show further that

$$\left(\frac{2}{\gamma + 1}\right) u_L^2 + u_L(u_R - u_L) - \left(\frac{2\gamma}{\gamma + 1}\right)\frac{p_L}{\rho_L} = 0,$$ (2.112)

and that

$$\left(\frac{2}{\gamma + 1}\right)\frac{\rho_L u_L^2}{p_L} - \frac{p_R}{p_L} - \frac{\gamma - 1}{\gamma + 1} = 0.$$ (2.113)

Now consider a plane shock lying in the plane $x = X(t) < 0$ and impinging on a stationary solid wall at $x = 0$. Prior to the passage of the shock the gas is at rest with pressure p_0 and density ρ_0. As the shock moves towards the wall with steady velocity $dX/dt = U_+ > 0$, the gas behind the shock has velocity u_s, where $0 < u_s < U_+$, pressure p_s and density ρ_s. After the shock has rebounded from the wall it moves with velocity $dX/dt = -U_- < 0$ into the already once-shocked gas. The gas between the shock and the wall is now stationary and has pressure p_1 and density ρ_1. Use eq. (2.112) on both the pre- and post-rebound configurations to show that $(u_s + U_-)$ and $(u_s - U_+)$ both satisfy the same quadratic equation. Deduce that

$$(u_s - U_+)(u_s + U_-) = -\gamma p_s/\rho_s.$$ (2.114)

Similarly apply eq. (2.113) to both the pre- and post-rebound configurations and hence, using eq. (2.114), obtain the following relationship:

$$\left(\frac{2\gamma}{\gamma + 1}\right)^2 = \left(\frac{p_0}{p_s} + \frac{\gamma - 1}{\gamma + 1}\right)\left(\frac{p_1}{p_s} + \frac{\gamma - 1}{\gamma + 1}\right).$$ (2.115)

In the case of a strong shock ($p_0 \ll p_s$), show that

$$\frac{p_1}{p_s} = \frac{3\gamma - 1}{\gamma - 1}.$$ (2.116)

(See Billingham & King (2000).)

2.4.7 At time $t = 0$, an infinite tube contains gas with uniform density and sound speed c_0 in the range $x > 0$ and has a stationary piston at $x = 0$. For $t > 0$ the piston moves subsonically with constant velocity $-U$, where $U > 0$.

Assume that all the C_- characteristics which originate on the line $t = 0, x > 0$ terminate on the piston. Deduce that two C_+ characteristics emanate from the position $x = 0, t = 0$, one of which ($x = c_0 t$) represents the front of a rarefaction wave and the other of which ($x = [c_0 - \frac{1}{2}(\gamma + 1)U]t$) represents the back. Show that all other C_+ characteristics are parallel to one or other of these two. Hence sketch all the characteristics of the flow in the (x, t)-plane.

Show that within the rarefaction wave the solution takes the following form:

$$x = [u + c(u)]t, \qquad (2.117)$$

and deduce that at that point

$$| u | = \frac{2}{\gamma + 1}(c_0 - x/t). \qquad (2.118)$$

Use this information to sketch the velocity $u(x, t)$ and the density $\rho(x, t)$ at some later time $t > 0$. What happens if $U > 2c_0/(\gamma - 1)$? (See Landau & Lifschitz (1959, Chap. X) and Zel'dovich & Raizer (1967, Chap. 1.).

3

Spherically symmetric flows

Many astrophysical phenomena are approximately spherically symmetric. Stars are an obvious example, provided that we can neglect the effects of rotation and magnetic fields. Then the stellar wind is essentially a steady spherical outflow. But spherical symmetry is often a good description of cases where a fluid with a large lengthscale and little angular momentum is affected by the presence of a smaller object. An important example of this is gas falling on to a small mass embedded in it. This may describe a star accreting gas from the interstellar medium, or the capture of gas by the nucleus of an active galaxy. If conditions far from the central object change only slowly we can assume steady inflow or outflow.

3.1 Steady inflow/outflow

We consider a steady, spherically symmetric flow, centred on the origin at which there is a point mass, M, with gravitational potential given by

$$\Phi(r) = -\frac{GM}{r},$$
(3.1)

where r is the spherical radius. To keep things simple, we neglect thermal processes and assume that the fluid is isentropic. The density $\rho(r)$ and pressure $p(r)$ are just functions of radius, r, and the velocity is radial with magnitude $u(r)$, which also depends only on the radius r. The mass conservation equation, eq. (1.4), becomes

$$\frac{\partial \rho}{\partial t} + \frac{1}{r^2}\frac{\partial}{\partial r}(\rho r^2 u) = 0.$$
(3.2)

Since the situation is steady, we can set $\partial \rho / \partial t = 0$ and then integrate once with respect to r. The equation then tells us the obvious result that the mass flux, $\dot{M}$, through each spherical shell, radius r, is the same. Thus we can write

$$4\pi r^2 \rho u = 4\pi A,$$
(3.3)

44

where A is a constant. For an outflow, or wind, the mass loss rate is $\dot{M} = 4\pi A$, with $A > 0$ and $u > 0$. For an inflow, the mass accretion rate is $\dot{M} = -4\pi A$, with $A < 0$ and $u < 0$.

For an isentropic flow we have seen that $p = K\rho^\gamma$, where K is a constant and γ is the ratio of specific heats. For a monatomic gas, $\gamma = 5/3$ and in general $1 \leq \gamma \leq 5/3$. Then

$$\int \frac{\mathrm{d}p}{\rho} = \left(\frac{\gamma}{\gamma - 1}\right) \frac{p}{\rho}, \tag{3.4}$$

and Bernoulli's equation, eq. (1.57), becomes

$$-\frac{GM}{r} + \frac{1}{2}u^2 + \left(\frac{\gamma}{\gamma - 1}\right) \frac{p}{\rho} = B, \tag{3.5}$$

where B is a constant.

Note that what we have done here is to replace the differential equations describing mass and momentum conservation by integral relations. This is possible because of the extreme simplicity of the case we are considering. In practice, in a more complex case it is often a sensible approach (often the only practicable one) to cast the problem as a set of differential equations and to then integrate these equations numerically. In the present simple case we can describe the properties of the flow by geometrical methods, starting from the integral relations.

3.1.1 Bondi accretion

To be specific, we consider the problem of determining the mass accretion rate, $\dot{M}$, on to a gravitating point mass, M, from an external (non-gravitating) medium, which at large distance is at rest and has uniform density ρ_∞ and uniform pressure p_∞. This problem and its solution were first described by Bondi (1952). To clarify the algebra, we replace u by $v = -u > 0$, which is the inward radial flow speed.

We start by asking what we might expect from the physics of this problem. First, we expect accretion to occur on to the central point mass. Second, as we move inwards from infinite radius, we expect the velocity to increase, and we expect the density (and hence also the sound speed in an isentropic flow) to increase. We denote the sound speed at infinity by c_∞, where $c_\infty^2 = \gamma p_\infty / \rho_\infty$. Then, the physical quantities defined by the problem are the central mass M and the sound speed at infinity c_∞. These, together with the gravitational constant G can be used to define a radius $r_G = GM/c_\infty^2$. What might be the physical relevance of this radius? The radius corresponds roughly to the radius at which the escape velocity from the point mass is equal to the sound speed at infinity. Thus, loosely speaking, we would expect matter well outside this radius to not 'know' about the point mass, i.e. not to be strongly influenced by its presence. Outside this radius we expect pressure forces within the fluid to play a significant role. That is, we expect the flow to be subsonic. In contrast, we expect matter well within this radius to be captured by the point mass. We expect the central gravitational pull to overwhelm the pressure forces, and therefore we expect matter to be falling freely (and supersonically, since pressure is unimportant) on to the centre. Close to the centre we might expect the velocity to approach the free fall velocity $v \sim (2GM/r)^{1/2}$.

Let us now look at the problem in more detail. We work in the (c_s, v)-plane, where the sound speed c_s is given as usual by $c_s^2 = \gamma p/\rho$. By considering eq. (3.5) in the limit $r \to \infty$, we see that in this case

$$B = \frac{c_\infty^2}{\gamma - 1}, \tag{3.6}$$

and thus that the equation becomes a relationship between v and c_s in the following form:

$$\frac{1}{2}v^2 + \frac{c_s^2}{\gamma - 1} = \frac{c_\infty^2}{\gamma - 1} + \frac{GM}{r}. \tag{3.7}$$

From a geometrical point of view, this relationship describes a set of similar ellipses in the (c_s, v)-plane, centred on the origin, whose size increases as r decreases.

Using the relationship between density and sound speed for this isentropic fluid in the form $c_s^2 = \gamma K \rho^{(\gamma - 1)}$, we can write eq. (3.3) as follows:

$$v = \left(\frac{-A}{\rho_\infty r^2} \right) \left(\frac{c_\infty}{c_s} \right)^{(2/\gamma - 1)}. \tag{3.8}$$

From a geometrical point of view, assuming $\gamma > 1$, this relationship describes a set of similar hyperbolae in the (c_s, v)-plane, with asymptotes being the two axes $v = 0$ and $c_s = 0$, and whose size also increases with decreasing r.

At each radius r the solution of the flow equations, i.e. the values of $c_s(r)$ and $v(r)$, are given by the point(s) at which the two curves eqs. (3.8) and (3.7) intersect in the (c_s, v)-plane.

In Fig. 3.1(a) we consider the picture at very large radius $r = \infty$. Here the ellipse given by eq. (3.7) intersects the c_s-axis at point P where $c_s = c_\infty$, and intersects the v-axis at point Q where $\frac{1}{2}v^2 = c_\infty^2/(\gamma - 1)$. The hyperbola given by the limits of eq. (3.8) as $r \to \infty$ is given by the two axes $v = 0$ and $c_s = 0$. These curves intersect at the two points P and Q. The point we are interested in for the Bondi accretion problem is the point $(c_\infty, 0)$, marked P in Fig. 3.1, corresponding to zero velocity and finite sound speed at infinity. (We discuss the other point, marked Q in Fig. 3.1, in Section 3.1.2.)

Now, as r decreases a little, as shown in Fig. 3.1 (b), both curves shift to larger values of c_s and v. Therefore the intersection points move P and Q to larger values of c_s and of v. Thus for the solution we are interested in as r decreases and the fluid flows inwards, both sound speed and velocity increase, as expected.

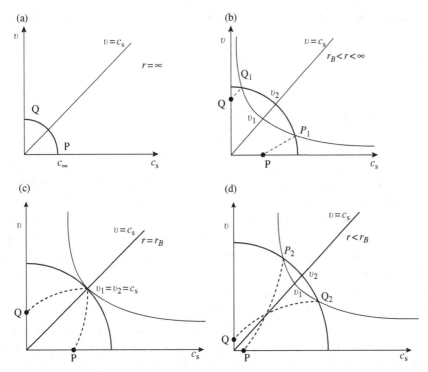

Fig. 3.1. The Bondi problem viewed in the (c_s, v)-plane. The ellipses are curves of constant accretion rate, and the hyperbolae are the Bernoulli constant, each specified by boundary conditions at infinity. The pictures at (a) large radius $r = \infty$ from the central accretor; (b) $r_B < r < \infty$; (c) $r = r_B$; (d) $r < r_B$.

The question now is: What happens next? We see from eqs. (3.8) and (3.7) that as $r \to 0$ both sets of curves continue to increase in size indefinitely. What happens to the intersection points? There are clearly three possibilities.

(i) At some radius, the curves may cease to intersect. This would mean that just inside this radius there is no mathematical solution to the equations. This is clearly untenable from a physical point of view. But what could we do if this happened? The answer is to note that in order to draw the curves we have had to assume a value for the accretion rate $\dot{M}$. To be specific, we see that from eq. (3.8) the size of the hyperbola in the v-direction depends on the magnitude of the constant A, i.e. on the accretion rate $\dot{M}$. Thus if at some radius the curves cease to continue to intersect, we could make them continue to intersect by adjusting the assumed accretion rate downwards. This tells us that adjusting the accretion rate is the key to finding a solution to the flow problem.

The point at which the curves almost stop intersecting is the same as the point at which the two points of intersection coincide. This would occur at a point in the

(c_s, v)-plane when the two curves described by eqs. (3.8) and (3.7) have the same slope.

The slope of eq. (3.7) at fixed radius r is given by

$$v\frac{dv}{dc_s} = -\left(\frac{2}{\gamma - 1}\right)c_s,\tag{3.9}$$

and the slope of eq. (3.8) at fixed radius r is given by

$$\frac{1}{v}\frac{dv}{dc_s} = -\left(\frac{2}{\gamma - 1}\right)\frac{1}{c_s}.\tag{3.10}$$

Thus the slopes are equal when $v = \pm c_s$, i.e. when the flow velocity is transonic. This leads us to the second possibility.

(ii) The curves intersect at two points at all radii. Then the solution with the correct boundary conditions at large radius would be given by the curve in the (c_s, v)-plane obtained by tracing the set of intersection points as r decreases, starting at the point $(c_\infty, 0)$ corresponding to $r = \infty$. This would then imply that the flow is subsonic at all radii. We can see this because at small radii, as $r \to 0$, if $v \ll c_s$ eq. (3.7) implies $c_s \propto r^{1/2}$. From this we find that eq. (3.8) implies $v/c_s \propto r^{(5-3\gamma)}$ and thus $v/c_s \to 0$ as $r \to 0$, given that $\gamma < 5/3$. Thus, while this is an acceptable solution from both mathematical and physical points of view, it corresponds to a pressure-supported atmosphere slowly settling on to the central point mass. This is not the solution we are looking for. Thus we need to consider the third possibility.

(iii) The curves intersect at two points at all radii, except for one radius at which they just touch. We shall call this radius the Bondi radius, r_B. At this radius we see that by equating the slopes in eqs. (3.9) and (3.10) the two solutions intersect at a point where $v = c_s$. This gives us the possibility of a global solution to the problem which is subsonic at large radii and is supersonic at small radii. From the discussion above we expect to have to choose the accretion rate exactly to give this solution. In other words, insistence on this solution being the physically sensible one determines the accretion rate. Let us see how this comes about.

We know that if the two curves given by eqs. (3.8) and (3.7) touch they do so at a point on the line $v = c_s$ (see Fig. 3.1 (c)). So we define as $v_1(r)$ the value of v as a function of r at which the curves in eq. (3.8) and $v = c_s$ intersect (see Fig. 3.1(b)). Similarly, we define as $v_2(r)$ the value of v as a function of r at which the curves in eq. (3.7) and $v = c_s$ intersect (see Fig. 3.1). Thus v_1 and v_2 are given by the following equations:

$$v_1(r)^{\gamma+1/\gamma-1} = \left(\frac{A}{\rho_\infty}\right)c_\infty^{2/(\gamma-1)}\frac{1}{r^2},\tag{3.11}$$

and

$$v_2^2(r) = \frac{2(\gamma - 1)}{\gamma + 1}\left[\frac{c_\infty^2}{\gamma - 1} + \frac{GM}{r}\right].\tag{3.12}$$

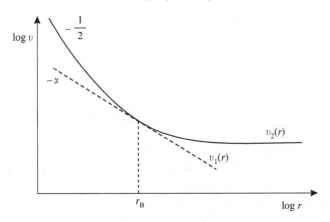

Fig. 3.2. Logarithmic dependence on radius of the two sonic point velocities v_1, v_2 in the Bondi problem.

At large radius, $r \to \infty$, we can see from the definitions (or from Fig. 3.2) that $v_1 \to 0$ whereas v_2 is finite. Thus, at large radii, $v_1 < v_2$. Indeed, from Fig. 3.1, we can see that for there to be two (or one) solutions at each radius, we require $v_1 \leq v_2$ at all radii, with equality only at one radius, r_B. At the Bondi radius, $r = r_B$, these two curves must just touch. In other words, the equation $v_1(r) = v_2(r)$ must have only one solution for r. Consider first the curve $v_2(r)$. We have noted that at small radii $v_2(r) \propto r^{-1/2}$, and we see that at large radii $v_2(r) \propto$ constant. We sketch this on logarithmic axes in Fig. 3.2. However, at all radii we see that $v_1(r) \propto r^{-\alpha}$, where $\alpha = 2(\gamma - 1)/(\gamma + 1)$. We note that $\alpha > 0$ for $\gamma > 1$. This too is drawn on logarithmic axes in Fig. 3.2. It is immediately evident that for these two curves to touch we require $0 < \alpha < 1/2$. A little algebra shows that this implies $1 < \gamma < 5/3$. In order to make them just touch, we see from eq. (3.11) that we must choose a particular value of A, i.e a particular value of the accretion rate $\dot{M}$. This is what we expected.

We can find the radius r_B at which these two curves touch by setting the slopes of the two curves equal. Thus,

$$\frac{d \ln v_2}{d \ln r} = -\alpha. \tag{3.13}$$

From this we find that

$$r_B = \frac{5 - 3\gamma}{4} \frac{GM}{c_\infty^2}. \tag{3.14}$$

Substituting this into eq. (3.12) we find that at this point

$$v_2^2 = \frac{GM}{2r_B}. \tag{3.15}$$

For the curves to touch, we require them to have not only the same slope, but also the same values of v. Thus at r_B we must set $v_1(r_B) = v_2(r_B)$. By doing this we determine the accretion rate. This is known as the Bondi accretion rate, $\dot{M}_B$, given by

$$\dot{M}_B = \left(\frac{2}{5 - 3\gamma}\right)^{\frac{\gamma+1}{2(\gamma-1)}} \cdot 4\pi r_B^2 \rho_\infty c_\infty. \qquad (3.16)$$

We note that this result makes physical sense. We argued earlier that some radius $r_G \sim GM/c_\infty^2$ would play an important physical role, and find that this is given by the Bondi radius r_B. To order of magnitude the accretion rate we get is simply the value we obtain from a flow through the sphere at $r = r_B$, at speed c_∞ and with density ρ_∞. This is what we would guess from the start. The main result of our analysis has been to obtain the constant of proportionality.

From Fig. 3.1 we see that at radii r_B the two solutions represented by the points P and Q interchange. The accretion solution is represented by point P and becomes supersonic inside r_B.

We also note that steady, spherically symmetric accretion of an isentropic fluid onto a point mass (Bondi accretion) can only occur if $1 < \gamma < 5/3$. As $\gamma \to 5/3$, we see that $r_B \to 0$ and $\dot{M} \to 0$. For values of γ larger than $5/3$ it is possible formally to set up a static atmosphere filling the whole of space, extending all the way from the point mass at $r = 0$ to infinite radius.

3.1.2 Steady outflow

Essentially all of the above analysis applies also to the situation in which the flow is outwards, with velocity $u(r) > 0$, and with outwardly directed mass loss rate $\dot{M}$. This corresponds to an isentropic stellar wind. The relations given by eqs. (3.8) and (3.7) are the same, except that $v(r)$ is replaced by $u(r)$. All we have to do is to follow the progress of the other point of intersection of the two curves, i.e. point Q in Fig. 3.1. This point starts at zero velocity at $r = 0$, and stays subsonic as far as the radius r_B. At that point the flow becomes supersonic, and remains so to large radius. We can see from Fig. 3.1 (d) that as $r \to \infty$, $u^2 \to 2c_\infty^2$ and $c_s \to 0$.

3.2 Explosion in a uniform medium

Here we consider the effects of an explosion at a point in a uniform medium. The first studies of this problem concerned explosions of nuclear bombs in the Earth's atmosphere. In astronomy this analysis is used to model the effects of the early stages of the explosion of a supernova in the interstellar medium.

We model the explosion as the instantaneous input of a fixed amount of energy E at time $t = 0$ at point $r = 0$ (in a spherical coordinate system) into a uniform background medium of density ρ_1. The flow is of course now a time-dependent one, and we therefore expect the velocity to be radially outwards and with magnitude $u(r, t)$. We expect the input of energy at such a large (formally infinite) rate to push the fluid outwards at such a large velocity (formally infinite initially) that it moves supersonically. In other words, we expect a shock wave to be generated at the origin at time $t = 0$ and to move radially outwards at some velocity $U_s(t)$. Initially, the shock will be 'strong' in the sense that the pressure p_2 behind the shock is much greater than the pressure in the pre-shocked gas, i.e. $p_2 \gg p_1$. This is equivalent to assuming that the sound speed in the unshocked gas is much less than the shock velocity.

With these assumptions, the only physical quantities relevant to the problem are the initial density ρ_1 and the explosion energy E. This comes from assuming that the explosion is essentially at a point, whereas in reality there must be an associated lengthscale. However, we expect the shock wave to spread so rapidly that its size is almost immediately much larger than the original explosion scale. All quantities appearing in the solution must therefore be combinations of ρ_1, E and the time t, and we can use dimensional analysis to find their dependences, at least up to dimensionless multiplicative functions.

In terms of mass M, length L and time T, our two defining quantities have dimensions as follows:

$$\rho_1 \propto [\mathrm{ML}^{-3}] \tag{3.17}$$

and

$$E \propto [\mathrm{ML}^2\mathrm{T}^{-2}]. \tag{3.18}$$

Dividing these two equations, we eliminate M to obtain

$$\frac{\rho_1}{E} \propto \frac{\mathrm{T}^2}{\mathrm{L}^5}. \tag{3.19}$$

Since the l.h.s. is a constant, we conclude that any radius relevant to the problem must vary as $r \propto t^{2/5}$. From this we can deduce two things. First, the shock radius $r_s(t)$ must move outwards as $r_s \propto t^{2/5}$. Second, all relevant physical quantities such as velocity $u(r, t)$, pressure $p(r, t)$ and density $\rho(r, t)$ depend on radius and time only through the combination $\xi \propto r/t^{2/5}$.

3.2.1 Shock conditions

To proceed we must discuss how the various physical quantities u, p and ρ vary across a strong shock. The jump conditions across a stationary shock are the Rankine–Hugoniot conditions (eqs. (2.69), (2.72) and (2.80)) derived in Chapter 2.

In Problem 2.4.3, we deduced expressions for the jump in velocity (u_2/u_1) and the jump in density (ρ_2/ρ_1) in terms of the Mach number of the shock $\mathcal{M}$. For a strong shock we take the limit $\mathcal{M} \to \infty$ and obtain, in the frame in which the shock is stationary,

$$\frac{u_2}{u_1} = \frac{\gamma - 1}{\gamma + 1} \tag{3.20}$$

and

$$\frac{\rho_2}{\rho_1} = \frac{\gamma + 1}{\gamma - 1}. \tag{3.21}$$

However, here we shall be working in the inertial frame, in which the shock has a velocity $U_s(t)$. In the inertial frame, the fluid is at rest until the shock arrives. Thus in the frame of the shock, the fluid enters the shock with speed $u_1 = U_s$. Similarly, in the frame of the shock, the fluid leaves the shock with speed u_2, and therefore in the inertial frame, after the shock has passed, the fluid has velocity $u_2' = U_s - u_2$. We deduce that

$$u_2' = \left(\frac{2}{\gamma + 1}\right) U_s. \tag{3.22}$$

The post-shock density is unchanged by a Galilean transformation, and therefore the post-shock density is simply $\rho_2' = \rho_2$. In the limit $p_2 \gg p_1$, the Rankine–Hugoniot relation (eq. (2.72)) which describes momentum conservation yields in the frame of the shock the following:

$$p_2 = \rho_1 u_1^2 - \rho_2 u_2^2. \tag{3.23}$$

After a little algebra, we can use the above relations to show that

$$p_2 = \frac{2}{\gamma + 1} \rho_1 U_s^2. \tag{3.24}$$

3.2.2 Similarity variables

We have already deduced that the flow inside the shock ($r < r_s$) depends on radius r and time t only in the combination $r/t^{2/5}$, so we define a dimensionless variable,

$$\xi = r \left(\frac{\rho_1}{Et^2}\right)^{1/5}, \tag{3.25}$$

called a *similarity variable*. At fixed time t, the value of ξ is just proportional to the radius r. Thus, in terms of ξ the shock is at some fixed value, say $\xi = \xi_s$. Then the shock radius is given by

$$r_s(t) = \xi_s (E/\rho_1)^{1/5} t^{2/5} \tag{3.26}$$

and the shock velocity, $U_s = dr_s/dt$, is given by

$$\frac{U_s}{r_s} = \frac{2}{5t}. \tag{3.27}$$

Obviously we would like to write all the physical quantities u, ρ and p in dimensional form multiplied by functions of the dimensionless similarity variable ξ. Outside the shock the gas is unaffected by the explosion, and the solution is simply the initial configuration. So for $r > r_s$ we have $u = 0$, $\rho = \rho_1$ and $p = p_1 \approx 0$. Thus we need only concern ourselves with the solution inside the shock, at radii $0 < r < r_s(t)$, or equivalently $0 < \xi < \xi_s$. Since physical quantities inside the shock depend only on ξ, we may write

$$\rho(r,t) = \rho_2 A(\xi), \tag{3.28}$$

or equivalently

$$\rho(r,t) = \left(\frac{\gamma + 1}{\gamma - 1}\right) \rho_1 A(\xi), \tag{3.29}$$

where $A(\xi)$ is a dimensionless function of ξ. The definition of ρ_2 requires $A(\xi_s) = 1$.

Similarly, since the only combination of our fundamental quantities with the dimensions of velocity is r/t, we may expect that

$$u(r,t) = \lambda_0 (r/t) V(\xi), \tag{3.30}$$

where λ_0 is some constant and $V(\xi)$ is a dimensionless function of ξ. Using eq. (3.27) we can rewrite this as follows:

$$u(r,t) = \lambda_0' (U_s/r_s) r V(\xi), \tag{3.31}$$

where λ_0' is some other constant. Motivated by the value for the velocity just inside the shock, given by eq. (3.22), we choose λ_0' such that

$$u(r,t) = \left(\frac{2}{\gamma + 1}\right) U_s \left(\frac{r}{r_s}\right) . V(\xi), \tag{3.32}$$

which implies that at $r = r_s$, where $\xi = \xi_s$, we have $u = u_2$ and $V(\xi_s) = 1$. Tidying this up, we obtain

$$u(r,t) = \frac{4}{5(\gamma + 1)} \left(\frac{r}{t}\right) V(\xi). \tag{3.33}$$

In a similar fashion, since pressure has dimensions of density times velocity squared we can write

$$p(r,t) = C_0 (r/t)^2 \rho_1 B(\xi), \tag{3.34}$$

where C_0 is some constant and $B(\xi)$ is a dimensionless function of ξ. As before, we try to make things simple by choosing the constant C_0 such that $B(\xi_s) = 1$. We can do this by writing

$$p(r,t) = p_2 (r/r_s)^2 B(\xi), \tag{3.35}$$

as then $p = p_2$ when $r = r_s$. Then, using the expressions derived above for p_2 and for r_s, we obtain

$$p(r,t) = \frac{8}{25(\gamma + 1)} \rho_1 \left(\frac{r}{t}\right)^2 B(\xi). \tag{3.36}$$

We have now achieved our aim of writing u, ρ and p in dimensional terms multiplied by the dimensionless functions $A(\xi)$, $V(\xi)$ and $B(\xi)$, respectively. We can now determine these three functions, and thus the detailed form of the solution, by inserting these expressions into the conservation equations.

3.2.3 The similarity (Taylor–Sedov) equations

In spherical symmetry the mass conservation equation is given by

$$\frac{\partial \rho}{\partial t} = \frac{1}{r^2}\frac{\partial}{\partial r}(\rho r^2 u). \tag{3.37}$$

Similarly the momentum conservation equation is given by

$$\frac{\partial u}{\partial t} + u\frac{\partial u}{\partial r} = -\frac{1}{\rho}\frac{\partial p}{\partial r}. \tag{3.38}$$

For the energy conservation equation we assume that the shocked fluid does not cool, so that each fluid element conserves its entropy, i.e. $DS/Dt = 0$. Then we have

$$\frac{\partial}{\partial t}\left[\rho\left(e + \frac{1}{2}u^2\right)\right] + \frac{1}{r^2}\frac{\partial}{\partial r}\left[r^2\rho u\left(e + \frac{p}{\rho} + \frac{1}{2}u^2\right)\right] = 0, \tag{3.39}$$

and we assume a perfect gas so that $e = p/(\gamma - 1)\rho$.

When making the substitution we use the definition of ξ (eq. (3.25)), and hence the derivative of ξ with respect to t at fixed r is given by

$$\left.\frac{\partial \xi}{\partial t}\right|_r = -\frac{2\xi}{5t}, \tag{3.40}$$

and the derivative of ξ with respect to r at fixed t is given by

$$\left.\frac{\partial \xi}{\partial r}\right|_t = \frac{\xi}{r}. \tag{3.41}$$

Thus, in the conservation equations, we use the following relationships:

$$\left.\frac{\partial}{\partial r}\right|_t = \frac{\xi}{r}\left.\frac{\partial}{\partial \xi}\right|_t, \tag{3.42}$$

and, using the similarity assumption that $(\partial/\partial t)_\xi = 0$,

$$\left.\frac{\partial}{\partial t}\right|_r = -\frac{2\xi}{5t}\left.\frac{\partial}{\partial \xi}\right|_t. \tag{3.43}$$

We need to take a little care when making these substitutions. For this reason, we derive one equation in some detail and leave the other two to the reader. We rewrite the conservation equation in the following form:

$$\frac{\partial \rho}{\partial t} + \frac{\partial}{\partial r}(\rho u) + \frac{2\rho u}{r} = 0. \tag{3.44}$$

Substituting for ρ using eq. (3.29) and using eq. (3.40), the first term on the l.h.s. is simply given by

$$\frac{\partial \rho}{\partial t} = \left(\frac{\gamma+1}{\gamma-1}\right)\rho_1\left(-\frac{2\xi}{5t}\frac{dA}{d\xi}\right). \tag{3.45}$$

Similarly, using eq. (3.33) also, the third term on the l.h.s. of eq. (3.44) is given by

$$\frac{2\rho u}{r} = \frac{8}{5(\gamma-1)}\rho_1\frac{AV}{t}. \tag{3.46}$$

A little more care is required when considering the second term on the l.h.s. Substituting for ρ using eq. (3.29) and for u using eq. (3.33), and also using eq. (3.42) we obtain

$$\frac{\partial}{\partial r}(\rho u) = \frac{\xi}{r}\frac{\partial}{\partial \xi}\bigg|_t\left\{\frac{\gamma+1}{\gamma-1}\rho_1 A\frac{4}{5(\gamma+1)}\left(\frac{r}{t}\right)V\right\}. \tag{3.47}$$

The catch here is that we have to take the derivative of the contents of the curly bracket with respect to ξ at *fixed* t, but the bracket contains both t *and* r. Before taking the derivative, we must therefore replace r by ξ and t using eq. (3.25). When we do this, the second term on the l.h.s. of eq. (3.44) becomes

$$\frac{\partial}{\partial r}(\rho u) = \frac{4\rho_1}{5(\gamma-1)}\cdot\frac{1}{t}\frac{d}{d\xi}(\xi A V). \tag{3.48}$$

We can now gather the three terms together. We note that they each have a factor of $1/t$, which we can cancel (if this did not happen it would either be a sign that we have made an algebraic mistake, or be a sign that the assumed similarity variable did not work in this case). We obtain finally the mass conservation equation in similarity form:

$$-\xi\frac{dA}{d\xi} + \frac{2}{\gamma+1}\left(3AV + \xi\frac{d}{d\xi}(AV)\right) = 0. \tag{3.49}$$

Similarly, though at greater length, we can obtain the momentum conservation equation in similarity form:

$$-V - \frac{2}{5}\xi\frac{dV}{d\xi} + \frac{4}{5(\gamma+1)}\left(V^2 + V\xi\frac{dV}{d\xi}\right) = -\frac{2}{5}\frac{\gamma-1}{\gamma+1}\frac{1}{A}\left(2B + \xi\frac{dB}{d\xi}\right), \tag{3.50}$$

and, at even greater length, we can obtain the energy equation in similarity form:

$$-2(B+AV^2) - \frac{2}{5}\xi\frac{d}{d\xi}(B+AV^2)$$

$$+ \frac{4}{5(\gamma+1)}\left(5V(\gamma B + AV^2) + \xi\frac{d}{d\xi}[V(\gamma B + AV^2)]\right) = 0. \tag{3.51}$$

These three equations, eqs. (3.49), (3.50) and (3.51), are called the Taylor–Sedov equations.

3.2.4 Solving the Taylor–Sedov equations

The Taylor–Sedov equations are three, non-linear, first-order, ordinary differential equations for A, B and V as functions of ξ. We therefore require three boundary conditions to define a solution. In one sense we have the boundary conditions already, since our initial definitions were set up so that at the shock, i.e. at $\xi = \xi_s$, we have $A(\xi_s) = B(\xi_s) = V(\xi_s) = 1$. Thus, if, for example, we turned to a computer to integrate these equations numerically, starting from $\xi = \xi_s$ and working our way inwards, the problem looks as good as solved. But there is a snag. We still do not know the value of ξ_s – that is, we do not know where to start our numerical integration! In some sense, ξ_s is an eigenvalue of the problem, and we shall have to find some means of determining its value at the same time as we solve the differential equations. This problem was solved numerically by Taylor (1959a, b) and analytically by Sedov (1959). We start by considering the energy equation:

$$\frac{\partial}{\partial t}\left[\rho\left(e + \frac{1}{2}u^2\right)\right] + \frac{1}{r^2}\frac{\partial}{\partial r}\left[r^2\rho u\left(e + \frac{p}{\rho} + \frac{1}{2}u^2\right)\right] = 0, \tag{3.52}$$

and we integrate both sides of this equation over all space, namely $\int_0^\infty 4\pi r^2\, dr$, using the perfect gas equation of state $e = p/(\gamma - 1)\rho$. The second term makes zero contribution because $ru = 0$ at both $r = 0$ and $r = \infty$. The perfect gas law $e = p/(\gamma - 1)\rho$ then requires

$$\frac{d}{dt}\int_0^\infty \left(\frac{p}{\gamma - 1} + \frac{1}{2}\rho u^2\right) 4\pi r^2\, dr = 0. \tag{3.53}$$

The velocity is zero outside the shock, so we can split this integral into two parts as follows:

$$\frac{d}{dt}\int_0^{r_s(t)} \left(\frac{p}{\gamma - 1} + \frac{1}{2}\rho u^2\right) 4\pi r^2\, dr + \frac{d}{dt}\int_{r_s(t)}^\infty \frac{p_1}{\gamma - 1} 4\pi r^2\, dr = 0. \tag{3.54}$$

We can neglect the second integral on the l.h.s. since we have assumed throughout that $p_1 \ll p_2$. The equation then states physically that the total energy within the shocked fluid is a constant, and this constant is of course just the input energy E. Thus,

$$\int_0^{r_s(t)} \left(\frac{p}{\gamma - 1} + \frac{1}{2}\rho u^2\right) 4\pi r^2\, dr = E. \tag{3.55}$$

We can now translate this expression into our similarity variables, which yields

$$\frac{32\pi}{25(\gamma^2 - 1)} \int_0^{\xi_s} [B(\xi) + A(\xi)V^2(\xi)]\xi^4\, d\xi = 1. \tag{3.56}$$

In passing, we note that the total thermal energy E_{th},

$$E_{th} = \frac{32\pi E}{25(\gamma^2 - 1)} \int_0^{\xi_s} [B(\xi)]\xi^4\, d\xi, \tag{3.57}$$

and the total kinetic energy E_{kin},

$$E_{\text{kin}} = \frac{32\pi E}{25(\gamma^2 - 1)} \int_0^{\xi_s} [A(\xi)V^2(\xi)]\xi^4 \, d\xi, \tag{3.58}$$

are separately conserved.

We can now solve the problem. If we start with a guess for the value ξ_s, we can (in principle) find solutions for $A(\xi)$, $B(\xi)$ and $V(\xi)$. We can then compute the integral on the l.h.s. of eq. (3.56) and see if, as it should, it equals unity. If it does not (and of course usually for some guessed value of ξ_s it does not), we then have to change our initial guess and repeat the procedure. Thus we can imagine setting up an iterative procedure which will determine the value of ξ_s.

In Fig. 3.3 we show the solution for $\rho/\rho_2, p/p_2$ and u/u_2 as functions of $r/r_s = \xi/\xi_s$, for the particular value $\gamma = 7/5$, which is appropriate for air. For this value of γ, we find $\xi_s = 1.033$. Note that for small r we have $\rho/\rho_2 \to 0$, but $p/p_2 \to$ constant. This means that the ratio of temperatures tends to infinity, i.e. the shock wave has a very strong heating effect.

As we mentioned earlier, the original use of this solution was in studying the effect of nuclear explosions on the atmosphere. The dimensional result $r_s \sim (Et/\rho_1)^{2/5}$ allowed early estimates of the yield E of these devices. In astrophysics, supernova explosions conform to the Taylor–Sedov solution until the shock front becomes large enough that the mass of swept-up interstellar medium begins to slow it down

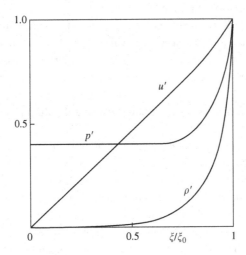

Fig. 3.3. The Taylor–Sedov similarity solution for a spherical blast wave in a uniform medium with $\gamma = 7/5$, appropriate for air. The quantities $\rho/\rho_2, p/p_2$ and u/u_2 (i.e. the density, pressure and velocity inside the shock in units of the pre-shock values) are plotted as functions of the similarity variable, or equivalently as functions of the radius in units of the shock radius.

(the 'snowplough' phase). More generally, the idea of a similarity solution is useful in situations where there is a globally conserved quantity (see Sedov (1959) and Zel'dovich & Raizer (1967)).

3.3 Further reading

The solution to the problem of steady accretion from a uniform medium was presented by Bondi (1952). The geometric solution given here is similar to that given in Zel'dovich & Novikov (1971, Chap. 13). Discussion of the similarity solution for a point explosion in a uniform medium is given in Landau & Lifshitz (1959, Chap. X) and in Shu (1992, Chap. 17). Further information about the use of similarity solutions in solving fluid dynamical problems is to be found in Zel'dovich & Raizer (1967, Chap. XII).

3.4 Problems

3.4.1 In a model of a rocket, a polytropic gas with adiabatic index $\gamma = 2$ flows steadily and adiabatically with velocity $v(x)$ along a smooth nozzle which has slowly varying cross-sectional area $S(x)$, where x measures the distance along the nozzle. At $x = 0$, S is very large so that the flow is very subsonic with sound speed c_0. Show that the velocity $v(x)$ and the sound speed $c(x)$ of the gas are related as follows:

$$\frac{1}{2}v^2 + c^2 = c_0^2, \tag{3.59}$$

and sketch this relationship in the (v, c)-plane.

As x increases, $S(x)$ decreases monotonically to a minimum value $S_{\min}$ at $x = x_{\min}$ and then increases to a very large value thereafter. The fluid velocity increases monotonically. Use mass conservation to plot another relationship in the (v, c)-plane for various values of $S(x)$, taking care to mark the curve corresponding to the value $S_{\min}$.

Show that at $x_{\min}$ $v = c_0\sqrt{2/3}$, and show that the fluid is eventually accelerated to a velocity of $v = c_0\sqrt{2}$.

Sketch the pressure profile $p(x)$ and explain how this process might be relevant to astrophysical jets.

(See Blandford & Rees (1974) and Königl (1982).)

3.4.2 A fluid flow with radial velocity $u(r)$ and density $\rho(r)$ represents the steady, spherically symmetric (Bondi) accretion of an isentropic fluid from a surrounding medium of uniform density ρ_∞ onto a gravitating point mass M centred at the origin ($r = 0$). Show that $u(r)$ and the adiabatic sound speed $c_s(r)$ obey the following equation:

$$\frac{1}{u}\frac{du}{dr} = \frac{1}{r} \cdot \frac{GM/r - 2c_s^2}{c_s^2 - u^2}, \tag{3.60}$$

and find the corresponding equation for dc_s/dr.

Show that at the radius r at which the flow is trans-sonic, the velocity is given by $u^2 = GM/2r$.

Verify that at large radii the equations permit a solution of the form $u \to 0$, $\rho \to \rho_\infty$ as $r \to \infty$.

Verify that at small radii the equations permit a solution of the form $u^2 \sim 2GM/r$ and $u^2 \gg c_s^2$ as $r \to 0$, provided that the ratio of specific heats, γ, is such that $\gamma < \gamma_{\text{crit}}$, where the value of γ_{crit} is to be determined.

A strongly magnetized star, radius R_*, has a dipole field of the form $\mathbf{B} = \nabla \psi$, where, using cylindrical polar coordinates (R, ϕ, z), we have $\psi \propto z/(R^2 + z^2)^{3/2}$. Bondi-like accretion is taking place in an axisymmetric fashion along the magnetic field lines onto a small circle of radius a ($\ll R_*$) at the magnetic pole. Show that the cross-sectional area of the flow varies with radius approximately according to $A = \pi a^2 (r/R_*)^3$ for $A \ll R_*^2$.

If the magnetic field is so strong that it remains exactly dipolar, show that it exerts no force on the fluid.

In this case show that the corresponding value of γ_{crit} that allows supersonic flow onto the star is $\gamma_{\text{crit}} = 7/5$. (See Pringle & Rees (1972).)

4

Stellar models and stellar oscillations

In the next few chapters we consider what happens if we perturb a stationary fluid configuration. The unperturbed configuration we have in mind is a body of fluid at rest in a stationary gravitational potential well. This potential might result from the self-gravity of the fluid itself, as for a star, or it might be produced by some external agency. An example of the latter case is the potential well produced by the dark matter component of a cluster of galaxies. The intracluster medium sits in this potential, without significantly contributing to it.

Studying perturbations in this way is important for a number of reasons. We can often use a linear analysis, and thus make things mathematically tractable. Working out when perturbations grow or not often provides us with a good idea of how a system will react, even to finite (non-infinitesimal) perturbations. In particular, we may be able to decide if the system is likely to react with drastic changes (instability), or settle down again to a state rather like its original one (stability). A system's reaction to perturbations also tells us a lot about its structure. Just as geophysicists learn about the Earth's interior by studying how it reacts to perturbations such as earthquakes, astronomers can use a similar technique (asteroseismology) to study the interior of stars.

4.1 Models of stars

To be specific we shall mainly consider perturbations to models of stars, although the results we find are generally applicable. We first consider briefly what is involved in making a stellar model. We assume that the star is spherically symmetric and is in hydrostatic equilibrium. This gives us our first two equations.

The star requires a pressure gradient to balance the force of gravity. Thus we must have

$$-\frac{1}{\rho}\frac{dp}{dr} = \frac{d\Phi}{dr},$$

(4.1)

where, because the star is self-gravitating, the gravitational potential is related to the stellar density by Poisson's equation:

$$\frac{1}{r^2}\frac{\mathrm{d}}{\mathrm{d}r}\left(r^2\frac{\mathrm{d}\Phi}{\mathrm{d}r}\right) = 4\pi G\rho. \tag{4.2}$$

These two equations can be written as two first-order differential equations by defining $m(r)$ as the mass within a sphere of radius r, i.e.

$$m(r) = \int_0^r 4\pi r^2\rho\,\mathrm{d}r, \tag{4.3}$$

or, in differential form,

$$\frac{\mathrm{d}m}{\mathrm{d}r} = 4\pi r^2\rho. \tag{4.4}$$

Then the equation of hydrostatic equilibrium becomes

$$\frac{\mathrm{d}p}{\mathrm{d}r} = -\frac{Gm\rho}{r^2}. \tag{4.5}$$

If the pressure, p, is known as a function of density, ρ, then these two equations are sufficient to determine the structure of the star. Examples of this case are white dwarfs and neutron stars, whose matter is completely degenerate. Otherwise we have to consider the thermal structure of the star, and we need two more equations. First we need energy conservation. If L_r is the outward energy flux through the shell at radius r, then in a steady state this energy flux must equal the total energy production rate within the sphere of radius r. For most stars the main energy production is through nuclear processes, and we may write

$$L_r = \int_0^r 4\pi r^2\rho\epsilon_{\mathrm{nuc}}\,\mathrm{d}r, \tag{4.6}$$

where, in standard notation, ϵ_{nuc} is the energy generation rate per unit mass through nuclear reactions. In differential form this equation becomes

$$\frac{\mathrm{d}L_r}{\mathrm{d}r} = 4\pi r^2\rho\epsilon_{\mathrm{nuc}}. \tag{4.7}$$

Next, we need to describe how the energy produced in the stellar interior makes its way to the surface. Typically we might expect the energy flux per unit area, i.e. $L_r/4\pi r^2$, to be proportional to the temperature gradient, $\mathrm{d}T/\mathrm{d}r$. Often the physical mechanism for transporting heat is conduction of heat by photons (radiative transfer), which requires

$$\frac{\mathrm{d}T}{\mathrm{d}r} = -\frac{3\kappa\rho}{4acT^3}\frac{L_r}{4\pi r^2}. \tag{4.8}$$

Here κ is the opacity, which describes the interaction of stellar matter and photons at the microscopic level, a is the radiation constant and c is the speed of light. This equation shows that the net result of the interaction between the stellar radiation field and matter, consisting of huge numbers of absorptions and re-emissions within extremely short distances, is a drift of radiant energy down the local temperature gradient, i.e. radially outwards.

To close the system of equations we require some understanding of the physical processes determining pressure p, opacity κ and energy generation rate ϵ_{nuc} as functions of local density ρ and temperature T. Under most non-degenerate conditions p is given as a function of ρ and T by the perfect gas law, and the opacity and nuclear energy generation rates are also known as functions of ρ and T. Then the set of four structure equations, which consist of four first-order differential equations for m, p, T and L_r, is well defined, and all we need are some boundary conditions. For a radiative star (i.e. one transporting energy by radiative transfer as assumed above, rather than by other processes such as convection) these are simple. At the centre, $r = 0$, we require (by definition) $m = L_r = 0$, while at the stellar surface, $r = R$, we require $p = 0$ and, to a good approximation, $T = 0$. The latter boundary condition neglects the effect of the star's atmosphere (the thin outer layer where photons can escape freely into space) on its detailed emission properties. However, this is usually unimportant, since the temperature here is far lower than in the stellar interior, and the true temperature profile is extremely close to that given by assuming $T = 0$ at $r = R$, except very close to the surface.

4.2 Perturbing the models

We give here the equations of stellar structure (eqs. (4.4), (4.5), (4.7) and (4.8)) for a radiative star in order to illustrate the kind of physical considerations which go into making models of stars. There are many excellent books on the intricacies of making models of stars, and on following the evolution of the models as the stars use up their nuclear fuel. We do not deal with this here. Instead, we ask: If we succeed in making a model of a star, is it stable? If we perturb it a little, what does it do? If the star is stable, we might expect it to oscillate, and perhaps for its oscillations to be damped. But if it is unstable, then what?

The theory of stellar oscillations is sufficiently complicated that one could fill a large book on that subject alone. Indeed, several authors already have. In general, computations detailed enough to compare with observation require one to retain all the complexity of the original stellar structure equations. This makes numerical solution unavoidable. In this book, however, we aim to explain the ideas and motivation behind various areas of astrophysical fluid dynamics, rather than to present the fine details, so we shall try to keep things simple. This means ditching some physical reality and hence the ability to compare results with observations in detail. However, we always keep enough of the relevant physics that we can establish basic physical concepts and principles.

One of the complications of the detailed theory is that stars are spherical, and spherical geometry introduces algebraic complexity. To keep things simple, and thus not obscure the physics with heavy and unilluminating algebra, we consider only 'square' or 'flat' stars, which we can treat in Cartesian geometry. Later on when we consider rotating stars, we shall have to consider cylinders. Clearly this procedure is not realistic in detail, but it does not do too much violence to the physics. Aside from losing some factors of π, the main effect we neglect is that of geometrical spreading in lowering density and compression in raising it. This can be important in some contexts. For example, geometrical spreading is the reason that isolated pressure perturbations do not always lead to shock waves in three dimensions, whereas this formally always happens in one space dimension, as we saw in Chapter 2. However, the small perturbations we consider generally do not produce large enough changes in radius for such effects to appear in stellar oscillations.

A further complication occurs if we try to perturb everything simultaneously. Of course this is what happens in reality, but if we attempt this mathematically it is hard to keep track of what causes what. Thus we shall proceed by only allowing some parts of the physical system to vary at a time, and then only when we want to understand the results of such a perturbation. Thus we shall ignore the effects of perturbing the energy equation, except when we wish to look specifically at the stabilizing or destabilizing effects of doing so. Similarly, we ignore perturbations to the self-gravity of the star (i.e. we shall assume that the gravitational potential is fixed) except when we wish to look at possible instabilities produced by self-gravity.

In summary, we proceed one step at a time, concentrating on what is happening in physical terms, and trying to avoid any algebra that may confuse the issue.

4.3 Eulerian and Lagrangian perturbations

In treating perturbations it is important to be very clear in how we compare the perturbed and unperturbed flows. In particular, we have to distinguish carefully between the Eulerian and Lagrangian pictures, where the first considers the fluid quantities at fixed points in space and the second keeps track of these quantities as a given fluid element moves. This entails a certain amount of initial algebra, but allows us to view fluid phenomena in the most physically revealing way.

We must first establish what we mean by a perturbation. We consider in general an unperturbed fluid flow in which the fluid particles follow trajectories of the form $\mathbf{r}_0(\mathbf{x}_0, t)$. Here $\mathbf{x}_0$ is a vector field which labels the fluid particles. The simplest way

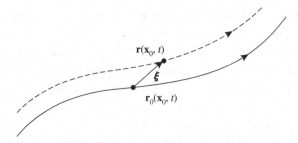

Fig. 4.1. The path of a particle (labelled $\mathbf{x}_0$) in the unperturbed flow is $\mathbf{r}_0(\mathbf{x}_0, t)$, shown by the solid line. The path of the same particle ($\mathbf{x}_0$) in the perturbed flow is $\mathbf{r}(\mathbf{x}_0, t)$ and is shown by the dashed line. The Lagrangian displacement is $\boldsymbol{\xi}(\mathbf{x}_0, t) = \mathbf{r}(\mathbf{x}_0, t) - \mathbf{r}_0(\mathbf{x}_0, t)$.

of thinking of it is as the particle position at time $t = 0$, i.e $\mathbf{x}_0 = \mathbf{r}_0$ at $t = 0$. Of course, if the unperturbed fluid is stationary, then $\mathbf{r}_0$ is independent of time t.

We then consider a small perturbation to the unperturbed flow. In the perturbed flow, the fluid particles, still labelled by the vector field $\mathbf{x}_0$, follow slightly different trajectories given by $\mathbf{r}(\mathbf{x}_0, t)$. Thus each fluid particle has been displaced by a small amount, $\boldsymbol{\xi}$, known as the Lagrangian displacement. Thus, we define

$$\boldsymbol{\xi}(\mathbf{x}_0, t) = \mathbf{r}(\mathbf{x}_0, t) - \mathbf{r}_0(\mathbf{x}_0, t) \tag{4.9}$$

(see Fig. 4.1). Although this looks a little complicated, we see that to each fluid particle, that is each $\mathbf{x}_0$, and for each time t, there corresponds a position of the particle $\mathbf{r}(\mathbf{x}_0, t)$ in the perturbed flow. Thus we can also regard $\boldsymbol{\xi}(\mathbf{r}, t)$ as a function simply of $\mathbf{r}$ and t. Then for each physical quantity f, e.g. pressure p or density ρ, we write f_0 for its value in the unperturbed flow and f for its value in the perturbed flow.

We now define the Eulerian perturbation of f to be the change in f as seen by an observer at some particular point $(\mathbf{r}, t)$. We denote the Eulerian perturbation by a prime, thus

$$f'(\mathbf{r}, t) = f(\mathbf{r}, t) - f_0(\mathbf{r}, t). \tag{4.10}$$

It is important to note that the Eulerian perturbation compares the properties of two different fluid particles. In addition, from the definition, it is evident that

$$\frac{\partial}{\partial t} f'(\mathbf{r}, t) = \frac{\partial}{\partial t} f(\mathbf{r}, t) - \frac{\partial}{\partial t} f_0(\mathbf{r}, t), \tag{4.11}$$

and therefore that $\partial/\partial t$ and $\prime$ commute. For the same reason ∇ and $\prime$ obviously commute. This is a useful property which we can use when considering the perturbed versions of the conservation equations.

It is often necessary to compare the properties of a particular fluid element in the two (perturbed and unperturbed) flows. This is the Lagrangian perturbation δf

defined by

$$\delta f(\mathbf{x}_0, t) = f(\mathbf{x}_0, t) - f_0(\mathbf{x}_0, t), \tag{4.12}$$

or equivalently

$$\delta f(\mathbf{r}, t) = f(\mathbf{r}, t) - f_0(\mathbf{r}_0, t), \tag{4.13}$$

where we recall that $\mathbf{r} = \mathbf{r}_0 + \boldsymbol{\xi}$.

The Eulerian and Lagrangian perturbations are therefore related by the exact expression

$$\delta f = f' + [f_0(\mathbf{r}, t) - f_0(\mathbf{r}_0, t)]. \tag{4.14}$$

Using a Taylor series we may write

$$f_0(\mathbf{r}, t) = f_0(\mathbf{r}_0, t) + \boldsymbol{\xi} \cdot \nabla f_0 + O(\xi^2). \tag{4.15}$$

Now to first-order in ξ we have $f \approx f_0$. This gives the basic result

$$\delta f = f' + \boldsymbol{\xi} \cdot \nabla f \tag{4.16}$$

to first order in ξ. We note that, since the Lagrangian perturbation δ comes from considering the same fluid element, it is evident that the procedure of forming the Lagrangian perturbation commutes with the Lagrangian derivative, that is

$$\frac{D}{Dt}(\delta f) = \delta \left(\frac{Df}{Dt} \right). \tag{4.17}$$

In addition, we need to add the warnings that $\prime$ does not commute with D/Dt and that δ commutes neither with $\partial/\partial t$ nor with ∇.

4.3.1 The perturbed velocity

In the above we have looked at expressions for the perturbations of scalar fluid quantities such as pressure and density. We need to take a little bit more care when considering perturbations of the velocity itself, and in particular the relations between these and the Lagrangian perturbation $\boldsymbol{\xi}$. We begin by noting that, for a fluid particle in the perturbed flow with trajectory $\mathbf{r}(t)$, the (perturbed) fluid velocity is given by

$$\mathbf{u}(\mathbf{r}, t) = \frac{D\mathbf{r}}{Dt}. \tag{4.18}$$

Similarly, for a fluid particle in the unperturbed flow with trajectory $\mathbf{r}_0(t)$, the (unperturbed) fluid velocity is given by

$$\mathbf{u}_0(\mathbf{r}_0, t) = \frac{D\mathbf{r}_0}{Dt}. \tag{4.19}$$

Then, taking the difference, we have by definition that

$$\delta \mathbf{u} = \mathbf{u}(\mathbf{r}, t) - \mathbf{u}_0(\mathbf{r}_0, t). \tag{4.20}$$

Noting that $\mathbf{r} = \mathbf{r}_0 + \boldsymbol{\xi}$, we see from the definition (eq. (1.19)) of the Lagrangian derivative that

$$\delta \mathbf{u} = \frac{\partial \boldsymbol{\xi}}{\partial t} + \mathbf{u} \cdot \nabla \boldsymbol{\xi} \tag{4.21}$$

to first order in $\boldsymbol{\xi}$.

Now, since by definition

$$\mathbf{u}' = \mathbf{u}(\mathbf{r}, t) - \mathbf{u}_0(\mathbf{r}, t), \tag{4.22}$$

and since, as we have shown (to first order)

$$\delta \mathbf{u} = \mathbf{u}' + \boldsymbol{\xi} \cdot \nabla \mathbf{u}, \tag{4.23}$$

we obtain finally the expression for the Eulerian velocity perturbation, valid to first order in $\boldsymbol{\xi}$:

$$\mathbf{u}' = \frac{\partial \boldsymbol{\xi}}{\partial t} + \mathbf{u} \cdot \nabla \boldsymbol{\xi} - \boldsymbol{\xi} \cdot \nabla \mathbf{u}. \tag{4.24}$$

If the unperturbed fluid is at rest, then these considerations are not really necessary and we have simply that $\delta \mathbf{u} = \mathbf{u}' = \mathbf{u} = d\boldsymbol{\xi}/dt = \partial \boldsymbol{\xi}/\partial t$.

4.4 Adiabatic perturbations – a variational principle

We are interested in deciding whether stars are stable against perturbations of various types. In principle we could explicitly solve the equations governing all possible perturbations of these types and see if any of the solutions grow without limit anywhere. Thus we need to know the *global* behaviour of the perturbations. A more convenient way of keeping track of this is to derive a *variational principle*. This involves an integral over the whole region of the fluid we are considering, and thus encapsulates the global information we require in a simple way. Deriving a variational principle involves some effort, but the payoff is that we can readily use it to decide the global stability of all perturbations in a straightforward manner.

We consider a fluid initially in hydrostatic equilibrium and subject to fixed gravity. In line with our philosophy of keeping things as simple as possible while retaining the essential physics, we assume a plane-parallel configuration in Cartesian coordinates. Then gravity is given by $\mathbf{g} = (0, 0, -g) = -\nabla \Phi$, where Φ is the gravitational potential, which we take to be fixed. Thus the unperturbed configuration satisfies following the equation:

$$\nabla p = -\rho \nabla \Phi. \tag{4.25}$$

In the initial configuration the pressure $p(z)$ and density $\rho(z)$ are just functions of z, and the velocity is zero. This implies that the perturbed velocity $\mathbf{u}$ is a first-order perturbed quantity, and we have $\mathbf{u} = \partial \boldsymbol{\xi}/\partial t$.

Then to first-order in perturbed quantities, the mass conservation equation,

$$\frac{\partial \rho}{\partial t} + \mathrm{div}(\rho \mathbf{u}) = 0, \tag{4.26}$$

becomes

$$\frac{\partial \rho'}{\partial t} + \mathrm{div}(\rho \mathbf{u}) = 0, \tag{4.27}$$

which we can integrate once with respect to time to yield

$$\rho' + \mathrm{div}(\rho \boldsymbol{\xi}) = 0. \tag{4.28}$$

Similarly, the momentum conservation equation,

$$\rho \frac{\partial \mathbf{u}}{\partial t} + \rho \mathbf{u} \cdot \nabla \mathbf{u} = \nabla p - \rho \nabla \Phi, \tag{4.29}$$

becomes

$$\rho \frac{\partial^2 \boldsymbol{\xi}}{\partial t^2} = -\nabla p' - \rho' \nabla \Phi. \tag{4.30}$$

Note here that since $\mathbf{u}$ is already a first-order quantity, the second term on the l.h.s. is second-order and so negligible, and also note that since gravity is fixed we can omit the term $-\rho \nabla \Phi'$.

We consider adiabatic perturbations. This means that each fluid element conserves its entropy as it moves, and further that the perturbation itself does not change the entropy of individual fluid elements. This is expressed most clearly by writing (in terms of the Lagrangian perturbation)

$$\delta \left(\frac{p}{\rho^\gamma} \right) = 0, \tag{4.31}$$

or equivalently

$$\delta p = \frac{\gamma p}{\rho} \delta \rho. \tag{4.32}$$

In terms of Eulerian perturbations, we have seen that we may write this as follows:

$$p' + \boldsymbol{\xi} \cdot \nabla p = \frac{\gamma p}{\rho} [\rho' + \boldsymbol{\xi} \cdot \nabla \rho]. \tag{4.33}$$

Thus we have an expression for the Eulerian pressure perturbation as follows:

$$p' = \frac{\gamma p}{\rho} \rho' - \boldsymbol{\xi} \cdot \left\{ \nabla p - \frac{\gamma p}{\rho} \nabla \rho \right\}. \tag{4.34}$$

We could now substitute eq. (4.28) for ρ' and eq. (4.34) for p' into eq. (4.30), and use eq. (4.25) describing the original equilibrium to give an equation solely in terms of the Lagrangian fluid perturbation $\boldsymbol{\xi}$. For any given initial perturbation $\boldsymbol{\xi}$, this equation would describe the time evolution of that perturbation. But it is more useful at this stage to look for more general properties of the perturbed flow.

Since the original configuration was independent of time t, we may replace the differential equation by an algebraic one, just as we did in Chapter 2, by taking the Fourier transform of the equation with respect to t. As we remarked earlier, this is equivalent to writing

$$\boldsymbol{\xi} \propto \exp(i\omega t). \tag{4.35}$$

When we do this, the full equation becomes

$$-\rho\omega^2\boldsymbol{\xi} = \nabla\left[\frac{\gamma p}{\rho}\mathrm{div}(\rho\boldsymbol{\xi})\right] + \nabla\left[\boldsymbol{\xi}\cdot\left(\nabla p - \frac{\gamma p}{\rho}\nabla\rho\right)\right] - \frac{1}{\rho}\nabla p\,\mathrm{div}(\rho\boldsymbol{\xi}). \tag{4.36}$$

Note that now $\boldsymbol{\xi}$ represents the Fourier transform of the Lagrangian perturbation with respect to time, and is in general a complex vector depending on ω and position $\mathbf{r}$.

We now look for general properties of this equation. To do this we multiply both sides by the complex conjugate quantity $\boldsymbol{\xi}^*$ and integrate over all space (or, in fact, since ρ vanishes outside the fluid, over all of the fluid).

The l.h.s. of the equation now becomes

$$\mathrm{LHS} = -\omega^2 I, \tag{4.37}$$

where I is a real, positive definite quantity given by

$$I = \int_V \rho\boldsymbol{\xi}^* \cdot \boldsymbol{\xi}\,\mathrm{d}V. \tag{4.38}$$

On the r.h.s. we first gather together the terms with a factor of γ. Simplifying, we obtain

$$\mathrm{RHS1} = \int_V \boldsymbol{\xi}^* \cdot \nabla[\gamma p\,\mathrm{div}\boldsymbol{\xi}]\,\mathrm{d}V. \tag{4.39}$$

We can now integrate this expression by parts. Assuming that the perturbation vanishes on the boundary, we obtain

$$\mathrm{RHS1} = -\int_V \gamma p\,\mathrm{div}\boldsymbol{\xi}\,\mathrm{div}\boldsymbol{\xi}^*\,\mathrm{d}V. \tag{4.40}$$

We can tidy up the rest of the terms on the r.h.s. (i.e. those independent of γ) to give us two quantities

$$A = \nabla(\boldsymbol{\xi} \cdot \nabla p) - \nabla p\,\mathrm{div}\boldsymbol{\xi} \tag{4.41}$$

and

$$B = -\nabla p(\boldsymbol{\xi} \cdot \nabla \ln \rho). \tag{4.42}$$

We take each of these in turn and apply the same procedure as before. From A, integrating the first term by parts and using the same boundary condition that the perturbation vanishes at the boundary, we obtain

$$\text{RHSA} = -\int_V dV\{\nabla p \cdot [\boldsymbol{\xi}\,\text{div}\boldsymbol{\xi}^* + \boldsymbol{\xi}^*\text{div}\boldsymbol{\xi}]\}. \tag{4.43}$$

And from B we obtain

$$\text{RHSB} = -\int_V dV\{(\boldsymbol{\xi}^* \cdot \nabla p)(\boldsymbol{\xi} \cdot \nabla \ln \rho)\}. \tag{4.44}$$

But since the original configuration was plane-parallel, so that

$$\nabla p = -|\nabla p|\hat{\mathbf{z}}, \tag{4.45}$$

and

$$\nabla \ln \rho = -|\nabla \ln \rho|\hat{\mathbf{z}}, \tag{4.46}$$

where $\hat{\mathbf{z}}$ is the unit vector in the z-direction, we can rewrite this in more symmetrical fashion as follows:

$$\text{RHSB} = \int_V dV\{(\nabla p \cdot \nabla \ln \rho)(\hat{\mathbf{z}} \cdot \boldsymbol{\xi}^*)(\hat{\mathbf{z}} \cdot \boldsymbol{\xi})\}. \tag{4.47}$$

Gathering all these expressions together, the transformed eq. (4.36) becomes simply

$$\omega^2 I = K, \tag{4.48}$$

where the quantity I is defined above and the quantity K is defined as follows:

$$K = \int_V dV\{\gamma p\,\text{div}\boldsymbol{\xi}\,\text{div}\boldsymbol{\xi}^* + \nabla p \cdot [\boldsymbol{\xi}\,\text{div}\boldsymbol{\xi}^* + \boldsymbol{\xi}^*\text{div}\boldsymbol{\xi}]$$
$$+ (\nabla_p \cdot \nabla \ln \rho)(\hat{\mathbf{z}} \cdot \boldsymbol{\xi}^*)(\hat{\mathbf{z}} \cdot \boldsymbol{\xi})\}. \tag{4.49}$$

4.4.1 Implications

The work going into producing eq. (4.49) now allows us to derive a number of results.

First, by construction, it is clear that both I and K are real quantities because both are unchanged by replacing $\boldsymbol{\xi} \leftrightarrow \boldsymbol{\xi}^*$. Further we can see that I is positive definite. This implies that ω^2 is real. Thus *either* $\omega^2 > 0$ and the perturbations are oscillatory, *or* $\omega^2 < 0$ and the perturbations are exponential, with one growing, and hence unstable, solution.

Second, since

$$\omega^2 = \frac{K[\xi]}{I[\xi]}, \tag{4.50}$$

we can see that this is a Sturm–Liouville problem. There is an extensive theory of such problems, treated at length in books on differential equations. In particular, one can always rewrite the problem as a variational principle . Here this means that the problem is equivalent to choosing a solution ξ_i which minimizes the quantity

$$F[\xi] = \frac{K[\xi]}{I[\xi]}. \tag{4.51}$$

The resulting quantity ξ_i is then an eigensolution to the original perturbation equations, and the corresponding eigenvalue is given by

$$\omega_i^2 = F[\xi_i]. \tag{4.52}$$

We also know that the eigenfunctions are orthogonal.

Third, since ω^2 is real, it follows that if we consider a continuous series of initial configurations which differ from each other by changing some parameter in a continuous fashion, then the change from stability to instability must occur when ω^2 passes through zero. That is, the 'exchange of stabilities' occurs when $\omega^2 = 0$.

4.4.2 Implication for stability

We have noted that, loosely, stability depends on whether ω^2 is greater or less than zero. But of course ω^2 depends on ξ. We now look in a little more detail to see how the above analysis can give us a criterion for the stability of the original configuration.

We started with an equation (eq. (4.30)) of the following form (after substitution):

$$\rho \frac{\partial^2 \xi}{\partial t^2} = L\xi, \tag{4.53}$$

where ξ is a function of $\mathbf{r}$ and t, and L is a time-independent linear operator. This means that

$$\frac{\partial}{\partial t} L\xi = L\frac{\partial \xi}{\partial t}. \tag{4.54}$$

In addition, a slight modification of the analysis at the beginning of this section, replacing ξ^* by η, shows that the operator L is symmetric, by which we mean that

$$\int_V \xi(L\eta) \, dV = \int_V \eta(L\xi) \, dV \tag{4.55}$$

for any two quantities $\xi(\mathbf{r}, t)$ and $\eta(\mathbf{r}, t)$.

We can now find criteria for stability and instability.

4.4.2.1 Stability

We show here that if the quantity

$$W(t) = \int_V \boldsymbol{\xi}(L\boldsymbol{\xi}) \, dV \tag{4.56}$$

is negative for *all* non-zero $\boldsymbol{\xi}$, then the configuration is stable. To keep the algebra simple, we assume that $\boldsymbol{\xi}$ is real. If this does not hold, then the same analysis results by setting one of the $\boldsymbol{\xi}$ on the r.h.s. to $\boldsymbol{\xi}^*$ and then taking the real part.

We note first that, since L is symmetric,

$$\frac{dW}{dt} = 2 \int_V \frac{\partial \boldsymbol{\xi}}{\partial t} \cdot (L\boldsymbol{\xi}) \, dV. \tag{4.57}$$

We then consider the quantity

$$K(t) = \frac{1}{2} \int_V \rho \frac{\partial \boldsymbol{\xi}}{\partial t} \cdot \frac{\partial \boldsymbol{\xi}}{\partial t} \, dV, \tag{4.58}$$

which represents the kinetic energy of the perturbation. It is clearly positive definite. Evidently,

$$\frac{dK}{dt} = \int_V \rho \frac{\partial^2 \boldsymbol{\xi}}{\partial t^2} \cdot \frac{\partial \boldsymbol{\xi}}{\partial t} \, dV, \tag{4.59}$$

and hence, using eqs. (4.53) and (4.57), we deduce that

$$\frac{dK}{dt} = \frac{1}{2} \frac{dW}{dt}. \tag{4.60}$$

Integrating with respect to t, this then implies that

$$2K(t) = W(t) + 2E, \tag{4.61}$$

where E is a constant. Since by assumption $W(t) < 0$ at all times t, we conclude that

$$K(t) < E \tag{4.62}$$

at all times t. Thus the kinetic energy of the perturbation is bounded above, and the configuration is stable.

4.4.2.2 Instability

We now show that if we can find some function $\boldsymbol{\eta}(\mathbf{r})$ such that

$$W_0 = \int_V \boldsymbol{\eta} \cdot (L\boldsymbol{\eta}) \, dV > 0, \tag{4.63}$$

then the original configuration is unstable.

We begin by defining $\omega_0 > 0$ by

$$W_0 = \omega_0^2 \int_V \boldsymbol{\eta} \cdot \boldsymbol{\eta} \, dV. \tag{4.64}$$

We then consider a perturbation of the fluid configuration such that at time $t = 0$ the displacement is given by

$$\boldsymbol{\xi}(\mathbf{r}, 0) = \boldsymbol{\eta}(\mathbf{r}) \tag{4.65}$$

and the perturbed velocity is given by

$$\frac{\partial \boldsymbol{\xi}}{\partial t} = \omega_0 \boldsymbol{\eta}. \tag{4.66}$$

Then, by construction, at time $t = 0$, and therefore for all time $t > 0$, we have $2K = W$ and hence $E = 0$.

We now consider the quantity $I(t)$ defined by

$$I(t) = \int_V \rho \boldsymbol{\xi} \cdot \boldsymbol{\xi} \, dV. \tag{4.67}$$

Differentiating twice, we find simply that

$$\frac{1}{2} \frac{d^2 I}{dt^2} = 2K + W, \tag{4.68}$$

and so in this case

$$\frac{d^2 I}{dt^2} = 8K. \tag{4.69}$$

We now use the Schwarz inequality in the following form:

$$\left(\int_V \rho \boldsymbol{\xi} \cdot \frac{\partial \boldsymbol{\xi}}{\partial t} \, dV \right)^2 \leq \left(\int_V \rho \boldsymbol{\xi} \cdot \boldsymbol{\xi} \, dV \right) \left(\int_V \rho \frac{\partial \boldsymbol{\xi}}{\partial t} \cdot \frac{\partial \boldsymbol{\xi}}{\partial t} \, dV \right) \tag{4.70}$$

to show that

$$\left(\frac{1}{2} \frac{dI}{dt} \right)^2 \leq I \times 2K. \tag{4.71}$$

We then substitute for K from eq. (4.69) to show that

$$\frac{d}{dt} \left(\frac{1}{I} \frac{dI}{dt} \right) \geq 0. \tag{4.72}$$

Because of the initial conditions we have chosen for the perturbation, we know that at $t = 0$, $I(dI/dt) = 2\omega_0 > 0$. We therefore conclude that, for all times $t \geq 0$,

$$\frac{1}{I} \frac{dI}{dt} \geq 2\omega_0, \tag{4.73}$$

and hence that

$$I(t) \geq I(0) \exp(2\omega_0 t), \tag{4.74}$$

and that the perturbation grows exponentially without limit.

4.4.2.3 Stability criterion

We conclude from all this that a necessary and sufficient condition for stability is that W defined in eq. (4.56) is negative for all vector fields $\boldsymbol{\xi}$. This is the global criterion we have been seeking, and we can use it to decide stability.

4.5 The Schwarzschild stability criterion

As an example we demonstrate how the above analysis gives a criterion for the stability of a fluid against convection.

We start from the equation describing the evolution of the perturbation, eq. (4.30):

$$\rho \frac{\partial^2 \boldsymbol{\xi}}{\partial t^2} = -\nabla p' - \rho' \nabla \Phi. \tag{4.75}$$

The equation which describes the adiabaticity of the perturbations, eq. (4.34) can be written as follows:

$$\rho' = \frac{\rho}{\gamma p} p' - \rho (\mathbf{A} \cdot \boldsymbol{\xi}), \tag{4.76}$$

where the quantity $\mathbf{A}$ is given by

$$\mathbf{A} = \frac{1}{\rho} \nabla \rho - \frac{1}{\gamma p} \nabla p. \tag{4.77}$$

We now recall that the r.h.s. of eq. (4.75) is just $L\boldsymbol{\xi}$. We then multiply this equation by $\boldsymbol{\xi}$, use eq. (4.76) to eliminate ρ', recall that $\nabla \Phi = -\nabla p / \rho$ and integrate over the volume of the fluid, to obtain

$$\int_V \boldsymbol{\xi} \cdot (L\boldsymbol{\xi}) \, dV = \int_V dV \left\{ p' \mathrm{div} \boldsymbol{\xi} + \frac{p'}{\gamma p} (\boldsymbol{\xi} \cdot \nabla p) - (\boldsymbol{\xi} \cdot \nabla p)(\mathbf{A} \cdot \boldsymbol{\xi}) \right\}, \tag{4.78}$$

where we have integrated the first term by parts and assumed that the perturbation vanishes on the boundary.

We can simplify this by using the mass conservation equation in the following form:

$$\mathrm{div} \boldsymbol{\xi} = -\frac{\rho'}{\rho} - \frac{1}{\rho} \boldsymbol{\xi} \cdot \nabla \rho, \tag{4.79}$$

to replace $\mathrm{div} \boldsymbol{\xi}$ in the first term on the r.h.s., and by then using eq. (4.76) to eliminate ρ'. Then two of the terms cancel and we obtain finally

$$\int_V \boldsymbol{\xi} \cdot (L\boldsymbol{\xi}) \, dV = -\int_V dV \left\{ \frac{(p')^2}{\gamma p} + (\boldsymbol{\xi} \cdot \nabla p)(\mathbf{A} \cdot \boldsymbol{\xi}) \right\}. \tag{4.80}$$

Now by the definition of $\mathbf{A}$ in eq. (4.77) we see that ∇p and $\mathbf{A}$ are either parallel or anti-parallel. We also note that, since the entropy $S \propto \ln(p/\rho^\gamma)$,

$$\mathbf{A} = -\frac{1}{\gamma}\nabla S. \tag{4.81}$$

Then, if entropy increases upwards, so that $\nabla p \cdot \mathbf{A} > 0$, we have that $W > 0$ for all $\boldsymbol{\xi}$, and therefore stability. On the other hand, if somewhere in the fluid entropy increases downwards, so that in some region of the fluid $\nabla p \cdot \mathbf{A} < 0$, then we can choose a vector function $\boldsymbol{\xi}$ which is sufficiently concentrated in that region that $W < 0$, and we have instability.

We conclude that the fluid is stable if and only if the entropy and pressure gradients are anti-parallel. Or, since in equilibrium gravity is anti-parallel to the pressure gradient, we conclude that the fluid is unstable if and only if entropy decreases downwards in the direction of gravity. This is the Schwarzschild criterion for convective instability. It implies, loosely speaking, that the fluid is stable if the hotter (higher entropy) fluid is on top and that convection sets in if the hotter (higher entropy) fluid is underneath. This corresponds directly to what we see when heating water from below, for example in a saucepan. In stars we often have stability against convection, even though the temperature decreases outwards. This is because the decrease in density ρ is sufficiently rapid that the specific entropy $\propto \ln(p\,\rho^{-\gamma})$ nevertheless increases outwards.

We can understand the Schwarzschild criterion by a simple physical argument, where we consider a fluid element displaced vertically upwards. Once it comes into pressure equilibrium with its new surroundings, we can see that it will fall back (stability against convection) provided that its density ρ is higher than its surroundings, i.e. provided that its entropy is lower than its surroundings. Assuming that the displacement was adiabatic, i.e. rapid enough for the element to conserve its entropy, this amounts to requiring entropy to decrease outwards in the unperturbed star, as the Schwarzschild criterion requires. In regions of stars where the Schwarzschild criterion fails, i.e. where entropy increases outwards, convection acts as an extremely efficient means of transporting energy out through the star. In fact, it is so efficient that the Schwarzschild criterion has only to be very slightly violated for the convection to carry the entire luminosity of the star. This is very convenient, for in many cases one can replace a detailed description of the convection with the simple condition that the entropy is constant, i.e. that the Schwarzschild criterion is only marginally violated.

4.6 Further reading

A description of the construction of models of stars is given in Clayton (1983, Chap. 6). Further clarification about the treatment of Eulerian and Lagrangian perturbations is given in Cox (1980, Chap. 5), who also discusses the Schwarzschild stability criterion (Chap. 17).

4.7 Problems

4.7.1 The equations governing the adiabatic perturbations, $\boldsymbol{\xi}e^{i\omega t}$, of a spherical star can be manipulated to yield the following expression:

$$I[\boldsymbol{\xi}]\omega^2 = K[\boldsymbol{\xi}], \tag{4.82}$$

where

$$I = \int \rho \boldsymbol{\xi}^* \cdot \boldsymbol{\xi} \, dV. \tag{4.83}$$

Show that taking account of the self-gravity of the perturbation gives a term of the form $-\rho\nabla\Phi'$ in the equations of motion. Show that combining this with the linearized version of Poisson's equation produces the following term:

$$G \int \int \frac{\text{div}[\rho(\mathbf{r})\boldsymbol{\xi}(\mathbf{r})]\text{div}'[\rho(\mathbf{r}')\boldsymbol{\xi}^*(\mathbf{r}')]}{|\mathbf{r} - \mathbf{r}'|} \, dV \, dV' \tag{4.84}$$

in $K[\boldsymbol{\xi}]$.

4.7.2 A spherically symmetric star, of radius R and mass M, undergoes small radial pulsations. The radial displacement vector is $\boldsymbol{\xi} = r\eta(r)\hat{\mathbf{r}}$. The pulsation frequency ω satisfies the equation

$$\omega^2 I = K, \tag{4.85}$$

where

$$I = \int_0^R \rho r^4 \eta^2 \, dr \tag{4.86}$$

and

$$K = \int_0^R \left\{ \gamma p r^4 \left(\frac{d\eta}{dr}\right)^2 - r^3\eta^2 \frac{d}{dr}[(3\gamma - 4)p] \right\} dr. \tag{4.87}$$

If

$$m(r) = \int_0^r 4\pi r^2 \rho \, dr, \tag{4.88}$$

show that, for a physically reasonable stellar density distribution $\rho(r)$,

$$\frac{m}{r^3} > \frac{M}{R^3}. \tag{4.89}$$

If γ is independent of radius, deduce that

$$\omega^2 > (3\gamma - 4)\frac{GM}{R^3}. \tag{4.90}$$

(See Cox (1980, Chap. 8).)

4.7.3 The equation of motion for a body of fluid in volume V, bounded by surface S, has the following form:

$$\rho \frac{d^2}{dt^2}(x_i) = -\frac{\partial}{\partial x_i}\left(p + \frac{1}{2\mu}B^2\right) - \rho\frac{\partial \Phi}{\partial x_i} + \frac{\partial}{\partial x_k}(B_i B_k). \tag{4.91}$$

Use this to prove the scalar virial theorem, which states that

$$\frac{1}{2}\frac{d^2 I}{dt^2} = 2T + W + 3(\gamma - 1)U + M + S, \tag{4.92}$$

where

$$I = \int_V \mathbf{r}\cdot\mathbf{r}\rho\,dV, \tag{4.93}$$

M is the magnetic energy defined by

$$M = \frac{1}{2}\int_V B^2\,dV, \tag{4.94}$$

T is the kinetic energy,

$$T = \frac{1}{2}\int_V \left(\frac{dx_i}{dt}\frac{dx_i}{dt}\right)\rho\,dV, \tag{4.95}$$

W is the gravitational energy ,

$$W = \frac{1}{2}\int_V \rho\Phi\,dV = -G\int_V \rho\mathbf{r}\cdot\nabla\Phi\,dV, \tag{4.96}$$

and U is the internal (thermal) energy,

$$U = \frac{1}{\gamma - 1}\int_V p\,dV, \tag{4.97}$$

and the surface integral S is given by

$$S = -\int_S \left(p + \frac{1}{2}B^2\right)\mathbf{r}\cdot d\mathbf{S} + \frac{1}{\mu}\int(\mathbf{r}\cdot\mathbf{B})\mathbf{B}\cdot d\mathbf{S}. \tag{4.98}$$

Assume $p = (\gamma - 1)\rho e$, with γ constant. (See Sturrock (1994, Chap. 12).)

4.7.4 Consider a fluid at rest, occupying volume V, with pressure distribution $p(\mathbf{r})$, density distribution $\rho(\mathbf{r})$ in a fixed gravitational field $\mathbf{g} = -\nabla\Phi(\mathbf{r})$ and permeated by a magnetic field $\mathbf{B}(\mathbf{r})$. Write down the equation describing the hydrostatic equilibrium.

The configuration undergoes a small oscillatory perturbation with displacement vector $\boldsymbol{\xi}(\mathbf{r})e^{i\omega t}$ and with div $\boldsymbol{\xi} = 0$. If the perturbation to the magnetic field is $\mathbf{b}(\mathbf{r})e^{i\omega t}$, show that

$$b_i = B_j\frac{\partial \xi_i}{\partial x_j} - \xi_j\frac{\partial B_i}{\partial x_j} \tag{4.99}$$

and deduce that div $\mathbf{b} = 0$.

Assuming (without proof) that all surface integrals vanish when integrating by parts show that

$$\omega^2 \int_V \rho\, \xi_i^* \xi_j \, dV = \int_V \xi_i^* \xi_j \frac{\partial^2}{\partial x_i \partial x_j} \left[p + \frac{1}{2} B^2 \right] dV + \int_V \rho\, \xi_i^* \xi_j \frac{\partial^2 \Phi}{\partial x_i \partial x_j} \, dV$$

$$+ \int_V \left(B_j \frac{\partial \xi_i^*}{\partial x_j} \right) \left(B_k \frac{\partial \xi_i}{\partial x_k} \right) dV$$

and hence that ω^2 is real.

Now consider a particular configuration in which the fluid is vertically stratified with $p(z)$ and $\rho(z)$ in a constant gravitational field $\mathbf{g} = (0, 0, -g)$, and with a horizontal magnetic field $\mathbf{B} = (B(z), 0, 0)$. Write down the equilibrium equation for this configuration.

By considering the perturbation $\boldsymbol{\xi} = (0, 0, \sin ky)$ in the above expression, comment on how stability depends on the sign of $\partial \rho / \partial z$.

Comment also on the stability properties of perturbations of the form $\boldsymbol{\xi} = (0, 0, \sin kx)$.

(A general variational principle for treating the stability of hydromagnetic systems is presented by Chandrasekhar (1961, Chap. IV).)

5

Stellar oscillations – waves in stratified media

In the previous chapter we showed that if a star is stable it oscillates about its equilibrium configuration when perturbed. To do this we looked at the global properties of the whole star. We showed that the star acts like an organ pipe in that it oscillates in a distinct set of modes with a distinct set of frequencies. But to investigate the details of the oscillation modes of the star, we need to look at the details of how the oscillations propagate through the star. In line with our philosophy expressed previously, we shall simplify matters by only considering flat 'stars', or equivalently we can think of the analysis as applying to a plane-parallel atmosphere, whose vertical thickness is small compared with the star's radius.

To understand the physics of the oscillations, we need to ask what the restoring forces are. That is, if we perturb the fluid, what tries to push it back to where it was? We have so far come across two types of restoring force in non-magnetic media, and we can expect both to operate in a star. They are pressure and buoyancy.

(i) Pressure. If we compress a fluid element, we increase its pressure, and this increase produces a restoring force. The resulting oscillations are sound waves, with local speed $c_s = \sqrt{\gamma p/\rho}$. Oscillation modes in which pressure is the main restoring force are called p-modes.

(ii) Buoyancy, or gravity. In a horizontally stratified fluid with gravity $\mathbf{g} = (0, 0, -g)$, we have seen that stability requires that the quantity A (see eq. (4.77)),

$$A = \frac{1}{\rho}\frac{d\rho}{dz} - \frac{1}{\gamma p}\frac{dp}{dz},\tag{5.1}$$

is negative. Such a fluid is stably stratified. If we perturb it in the vertical direction, gravity acts on the fluid elements to try to restore them to their original positions. It is clear that we do not need the fluid to be compressible for this to happen. A simple way to envisage these waves is to regard them as analogous to water waves on the surface of a pond or the sea. Then it is clear that a vertical displacement of the surface undergoes a restoring force due to gravity. From the physical quantities g and A, we can define a relevant frequency N, known as the Brunt–Väisälä frequency, given by

$$N^2 = -gA. \tag{5.2}$$

For these oscillations gravity is the main restoring force and they are known as g-modes. Note that N is real if and only if $A < 0$, i.e. the Schwarzschild criterion is satisfied. Thus g-modes propagate freely only in radiative regions of the star and are evanescent in convective ones.

We now investigate the properties of these modes of oscillation in more detail.

5.1 Waves in a plane-parallel atmosphere

We consider a horizontally stratified fluid with constant gravity $\mathbf{g} = (0, 0, -g)$. Thus in hydrostatic equilibrium the unperturbed values of $p(z)$ and $\rho(z)$ are related by

$$\frac{dp}{dz} = -\rho g. \tag{5.3}$$

We first perturb the equation of mass conservation, which is

$$\frac{\partial \rho}{\partial t} + \mathbf{u} \cdot \nabla \rho + \rho \operatorname{div} \mathbf{u} = 0, \tag{5.4}$$

to obtain, in terms of first-order Eulerian perturbed quantities,

$$\frac{\partial \rho'}{\partial t} + \mathbf{u} \cdot \nabla \rho + \rho \operatorname{div} \mathbf{u} = 0. \tag{5.5}$$

Similarly, we perturb the momentum equation,

$$\frac{\partial \mathbf{u}}{\partial t} + \mathbf{u} \cdot \nabla \mathbf{u} = -\frac{1}{\rho} \nabla p + \mathbf{g}, \tag{5.6}$$

to obtain

$$\frac{\partial \mathbf{u}}{\partial t} = \frac{\rho'}{\rho^2} \nabla p - \frac{1}{\rho} \nabla p'. \tag{5.7}$$

Note that we have assumed that gravity is fixed, so that $g' = 0$.

Because the unperturbed fluid is independent of time, and of the horizontal coordinates x and y, we can Fourier transform the equations with respect to these quantities, or equivalently we write all perturbed quantities in the form

$$p' \propto \exp\{i(\omega t - k_x x - k_y y)\}. \tag{5.8}$$

Note that now the perturbed quantities are still functions of the vertical coordinate z. Thus, for example, we regard the pressure perturbation as having the functional form $p' = p'(z; \omega, k_x, k_y)$. In addition, we shall write the components of the perturbed velocity in the following form:

$$\mathbf{u} = (u, v, w). \tag{5.9}$$

When we do this, eq. (5.5) becomes

$$i\omega\rho' + w\frac{d\rho}{dz} + \rho\left\{-ik_x u - ik_y v + \frac{dw}{dz}\right\} = 0. \tag{5.10}$$

Similarly, the three components of eq. (5.7) become

$$i\omega u = \frac{1}{\rho}ik_x p', \tag{5.11}$$

$$i\omega v = \frac{1}{\rho}ik_y p' \tag{5.12}$$

and

$$i\omega w = \frac{\rho'}{\rho^2}\frac{dp}{dz} - \frac{1}{\rho}\frac{dp'}{dz}. \tag{5.13}$$

We now need to provide a relationship between p' and ρ'. As before we shall assume that the oscillations are adiabatic so that energy is conserved. This implies that

$$\frac{\partial}{\partial t}\left(\frac{p}{\rho^\gamma}\right) + \mathbf{u}\cdot\nabla\left(\frac{p}{\rho^\gamma}\right) = 0. \tag{5.14}$$

The first-order Eulerian perturbation of this equation is given by

$$\frac{1}{\rho^\gamma}\frac{\partial p'}{\partial t} - \frac{\gamma p}{\rho}\frac{1}{\rho^\gamma}\frac{\partial\rho'}{\partial t} + w\frac{d}{dz}\left(\frac{p}{\rho^\gamma}\right) = 0. \tag{5.15}$$

Then applying Fourier transforms to this equation yields the following result:

$$i\omega p' - i\omega\frac{\gamma p}{\rho}\rho' + \rho^\gamma w\frac{d}{dz}\left(\frac{p}{\rho^\gamma}\right) = 0. \tag{5.16}$$

It is now more convenient to work in terms of the quantity $h' = p'/\rho$, which can be thought of as being related to the perturbation to the specific enthalpy $h = \int dp/\rho$. Then eq. (5.11) becomes

$$u = \frac{k_x h'}{\omega} \tag{5.17}$$

and eq. (5.12) becomes

$$v = \frac{k_y h'}{\omega}. \tag{5.18}$$

We use these to eliminate the horizontal velocity components u and v from the analysis. Substituting for u and v, eq. (5.10) becomes

$$i\omega\frac{\rho'}{\rho} + w\frac{1}{\rho}\frac{d\rho}{dz} + \frac{dw}{dz} - i\frac{(k_x^2 + k_y^2)h'}{\omega} = 0. \tag{5.19}$$

We replace p' with h' in eq. (5.13) to yield

$$i\omega w = \frac{\rho'}{\rho^2}\frac{dp}{dz} - \frac{dh'}{dz} - \frac{1}{\rho}\frac{d\rho}{dz}h' \tag{5.20}$$

and in eq. (5.16) to yield

$$\frac{\gamma p}{\rho}\frac{\rho'}{\rho} = h' - w\frac{i\rho^{\gamma-1}}{\omega}\frac{d}{dz}\left(\frac{p}{\rho^{\gamma}}\right). \tag{5.21}$$

We now have three equations for the three variables ρ'/ρ, h' and w. We note that the equations do not involve derivatives of ρ'/ρ. Thus we can use eq. (5.21) to substitute for ρ'/ρ in the other two equations.

Substituting in eq. (5.20), we obtain, after a little algebra,

$$\frac{\gamma p}{\rho}\frac{dh'}{dz} - \rho^{\gamma-1}\frac{d}{dz}\left(\frac{p}{\rho^{\gamma}}\right)h' = -iw\left\{\omega\frac{\gamma p}{\rho} - \frac{g\rho^{\gamma-1}}{\omega}\frac{d}{dz}\left(\frac{p}{\rho^{\gamma}}\right)\right\}, \tag{5.22}$$

where we have used the equilibrium condition that

$$g = -\frac{1}{\rho}\frac{dp}{dz}. \tag{5.23}$$

Then, using the fact that the expression for A can be rewritten in the following form:

$$A = -\frac{d}{dz}\left(\frac{p}{\rho^{\gamma}}\right) \times \frac{\rho^{\gamma}}{\gamma p}, \tag{5.24}$$

we may write the equation as follows:

$$\frac{dh'}{dz} + Ah' = -iw\left\{\omega + \frac{Ag}{\omega}\right\}. \tag{5.25}$$

Writing eq. (5.21) as

$$\frac{\rho'}{\rho} = \left(\frac{\rho}{\gamma p}\right)h' + \frac{iwA}{\omega}, \tag{5.26}$$

we substitute this into eq. (5.19) to obtain an equation for dw/dz in the following form:

$$\frac{dw}{dz} - \frac{g}{c_s^2}w + ih'\left[\frac{\omega}{c_s^2} - \frac{k_\perp^2}{\omega}\right] = 0. \tag{5.27}$$

Here we have used the notation that the horizontal component of the wave vector, $k_\perp$, is given by

$$k_\perp^2 = k_x^2 + k_y^2. \tag{5.28}$$

Equations (5.25) and (5.27) describe the oscillations. We can put them into neater form by recalling that $N^2 = -Ag$ and that the vertical displacement of a fluid element, ξ_z, is given by

$$w = i\omega\xi_z. \tag{5.29}$$

Using this, we write the equations in the final form:

$$\frac{d\xi_z}{dz} - \frac{g}{c_s^2}\xi_z + \frac{1}{c_s^2}\left[1 - \frac{k_\perp^2 c_s^2}{\omega^2}\right]h' = 0 \tag{5.30}$$

and

$$\frac{dh'}{dz} - \frac{N^2}{g}h' + (N^2 - \omega^2)\xi_z = 0. \tag{5.31}$$

5.1.1 Local analysis

For any plane-parallel distribution of density and pressure we can now use these two equations to find the oscillation frequencies, ω, and the structure of the oscillation modes. We do this for a particular set of oscillations in Section 5.2. Before we do that, however, it is instructive to look more generally at the kinds of oscillations that can propagate and to get a feel for the physics involved.

We undertake what is known as a local analysis and obtain a local dispersion relation, i.e. the relation between wavelength and frequency or wavenumber. We consider waves which have a vertical wavelength which is very small compared to the vertical scaleheight of the background distribution. To do this we set

$$\xi_z \propto \exp(ik_z z) \tag{5.32}$$

and

$$h' \propto \exp(ik_z z), \tag{5.33}$$

where $k_z \gg g/c_s^2$. In this approximation, eq. (5.30) becomes

$$ik_z \xi_z + \frac{1}{c_s^2}\left[1 - \frac{k_\perp^2 c_s^2}{\omega^2}\right]h' = 0 \tag{5.34}$$

and eq. (5.31) becomes

$$(N^2 - \omega^2)\xi_z + ik_z h' = 0. \tag{5.35}$$

These are two linear homogeneous equations, and so for a non-trivial solution we require the determinant of coefficients to vanish. This yields the following condition:

$$k_z^2 + \frac{1}{c_s^2}(N^2 - \omega^2)\left[1 - \frac{k_\perp^2 c_s^2}{\omega^2}\right] = 0. \tag{5.36}$$

This is the local dispersion relation, which can be written as a quadratic for ω^2 in the following form:

$$\omega^4 - (N^2 + k^2 c_s^2)\omega^2 + N^2 k_\perp^2 c_s^2 = 0, \tag{5.37}$$

where k is the full wavevector given by $k^2 = k_\perp^2 + k_z^2$.

In Fig. 5.1, we consider the local propagation properties of the oscillations in the $(k_\perp^2, \omega^2)$-plane. For mode propagation to take place we need *both* $k_\perp^2 \geq 0$ for horizontal propagation *and* $k_z^2 \geq 0$ for vertical propagation. In addition, we need

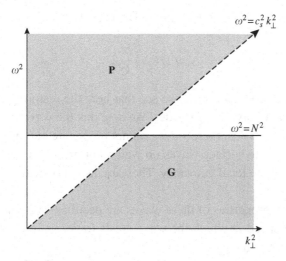

Fig. 5.1. The $(k_\perp^2, \omega^2)$-plane for stellar oscillations. The allowed regions for propagation of p- and g-modes are denoted 'P' and 'G', respectively.

$k^2 > 0$, so that at least one of $k_\perp^2$ and k_z^2 must be non-zero. The condition that $k_z^2 \geq 0$ requires one, and only one, of the brackets in eq. (5.36) to be negative. The areas in the $(k_\perp^2, \omega^2)$-plane in which propagation occurs are shaded in Fig. 5.1 and are marked 'P' and 'G'. They are separated from the areas in which propagation cannot occur by the lines on which $k_z^2 = 0$, on which the propagation is in the horizontal direction only. These lines are given by

$$\omega^2 = N^2,\tag{5.38}$$

which represents horizontally propagating gravity waves, and

$$\omega^2 = k_\perp^2 c_s^2,\tag{5.39}$$

which represents a sound wave which propagates horizontally with no z-dependence, known as the Lamb wave. The nature of the waves which propagate in region P can be seen by considering their behaviour in the limit of $\omega^2 \to \infty$ at fixed $k_\perp^2$. Taking this limit in eq. (5.37), or equivalently taking the limit $g \to 0$ or $N^2 \to 0$, we obtain

$$\omega^2 \approx k^2 c_s^2.\tag{5.40}$$

This, as we saw in Chapter 2, is the dispersion relation for sound waves. Thus the oscillations propagating in region P are the acoustic modes, or p-modes.

The nature of the waves propagating in the shaded region G follows from taking the limit in eq. (5.37) of low frequency $\omega^2 \to 0$, or by letting the sound speed be very large ($c_s^2 \to \infty$), which is equivalent to regarding the fluid as incompressible.

In this limit the dispersion relation, eq. (5.37), becomes

$$\omega^2 \approx N^2 \frac{k_\perp^2}{k^2}. \tag{5.41}$$

From the presence of the N^2 term, we see that here the restoring force is gravity. These are the gravity waves, or g-modes. We note that the waves need non-zero $k_\perp$ in order to propagate. Thus, these waves cannot propagate purely vertically. These waves are analogous to surface waves on water, but propagate within the body of the fluid because of the local buoyancy. They are also known as internal buoyancy waves.

The propagation properties of these waves are peculiar. The phase velocity , i.e. the direction and speed at which the wave crests travel, is in the direction of the wavevector and is given by

$$\mathbf{v_s} = \frac{\omega}{k} \hat{\mathbf{k}}. \tag{5.42}$$

However, the group velocity of the waves, which is the direction and speed at which the waves transport energy and momentum, is given by $\mathbf{v_g} = \partial\omega/\partial\mathbf{k}$. Using eq. (5.41) we find that for these waves

$$\mathbf{v_g} = \frac{\omega k_z}{k_\perp^2} \hat{\mathbf{k}} \wedge (\hat{\mathbf{k}} \wedge \hat{\mathbf{z}}). \tag{5.43}$$

From this we see that the group velocity is perpendicular to the phase velocity! Thus energy and momentum are carried parallel to the wave crests.

5.2 Vertical waves in a polytropic atmosphere

We now consider an example of how to use the wave equations we derived above to compute a set of oscillation modes in a particular distribution of fluid. We consider a fluid which is horizontally stratified, with constant gravity $\mathbf{g} = (0, 0, -g)$. We take the equation of state of the unperturbed fluid to be polytropic so that the unperturbed values of $p(z)$ and $\rho(z)$ are related by

$$p = K\rho^{1+1/m}, \tag{5.44}$$

where K is a constant, and $m > 0$ is also a constant, known as the polytropic index. We consider only those modes which propagate in the vertical direction. These are the equivalent of radial oscillation modes in a star. From our discussions above we already know that we expect these modes to be p-modes, because the g-modes cannot propagate vertically.

5.2.1 Equilibrium distribution

We must first determine the pressure and density distributions in the fluid. To do this we combine the equation of hydrostatic equilibrium,

$$\frac{dp}{dz} = -g\rho, \tag{5.45}$$

with the polytropic equation of state, eq. (5.44), to find that

$$\rho(z) = \left[\frac{g}{(m+1)K} \right]^m (-z)^m. \tag{5.46}$$

Here we note that this only describes the density in the region $z < 0$. We take the density to be zero in the region $z > 0$, so that the fluid represents a portion of a stellar atmosphere and has a surface which we have chosen to be at $z = 0$. For algebraic convenience we let $x = -z$, and thus the edge of the atmosphere is at $x = 0$ and it extends to positive x. We then find that $\rho \propto x^m$, pressure $p \propto x^{m+1}$ and temperature $T \propto x$. Thus the temperature increases linearly with the distance from the surface. We also find that the sound speed is given by

$$c_s^2 = \frac{\gamma g}{m+1} x \tag{5.47}$$

and that the buoyancy, $N^2/g = -A$, is given by

$$\frac{N^2}{g} = \frac{(\gamma-1)m - 1}{\gamma x}. \tag{5.48}$$

We note that the fluid has neutral buoyancy if it has uniform entropy, that is if $\gamma = 1 + 1/m$. And, as we expect from the Schwarzschild criterion (Section 4.5), stability (i.e. $N^2 \geq 0$) requires $\gamma \geq 1 + 1/m$.

5.2.2 The governing equation

To look at the details of vertically propagating waves, we set $k_\perp = 0$ in eqs. (5.30) and (5.31) to obtain

$$c_s^2 \frac{d\xi_z}{dz} - g\xi_z + h' = 0 \tag{5.49}$$

and

$$\frac{dh'}{dz} - \frac{N^2}{g} h' + (N^2 - \omega^2)\xi_z = 0. \tag{5.50}$$

We take eq. (5.49) and its derivative to obtain expressions for h' and dh'/dz in terms of ξ_z and its first and second derivatives. These we then substitute into eq. (5.50), and use eqs. (5.47) and (5.48), to obtain a second-order differential equation for ξ_z. Setting $x = -z$, we find the resulting equation:

$$c_s^2 \frac{d^2\xi_z}{dx^2} + \gamma g \frac{d\xi_z}{dx} + \omega^2 \xi_z = 0. \tag{5.51}$$

We note that from eq. (5.47) the coefficient of the first term $c_s^2 \propto x$. As we expected for vertically propagating waves, the buoyancy term involving N^2 has dropped out.

5.2.3 Solution of the equation

We have argued that these oscillation modes must just be p-modes. Thus we expect the character of the waves to be acoustic waves travelling in a medium in which the sound speed changes along the direction of propagation.[†] Rather than using the distance from the top of the atmosphere, x, as the independent variable, it then makes sense to use time of travel of acoustic waves from the top of the atmosphere, τ, where

$$\tau = \int_0^x \frac{dx}{c_s}. \tag{5.52}$$

Using eq. (5.47), we find that the relevant substitution is given by

$$x = \frac{1}{4}\left(\frac{\gamma g}{m+1}\right)\tau^2. \tag{5.53}$$

Using this in eq. (5.51), we find that the governing equation becomes

$$\frac{d^2\xi_z}{d\tau^2} + \frac{2m+1}{\tau}\frac{d\xi_z}{d\tau} + \omega^2\xi_z = 0. \tag{5.54}$$

The solution of this equation can be written in terms of Bessel functions:

$$\xi_z = \tau^{-m}\{C_1 J_m(\omega\tau) + C_2 J_{-m}(\omega\tau)\}, \tag{5.55}$$

where C_1 and C_2 are constants.

From the mathematical point of view, it makes sense to demand that ξ_z is finite at the surface $\tau = 0$. Recalling the property of Bessel functions that $J_m(Z) \sim Z^m$ as $Z \to 0$, we see that this requires $C_2 = 0$. From a physical point of view, the surface is defined by having zero pressure outside it. Once the surface is oscillating, we require that the pressure remains zero at the new position of the surface. Thus we require that the Lagrangian perturbation, $\delta p = p' + \xi_z(dp/dz)$, vanishes at $z = 0$. Note that this is not the same as demanding that p' vanish there! Noting that $dp/dz = -g\rho$, this condition becomes

$$h' - g\xi_z = 0 \tag{5.56}$$

[†] An exactly analogous problem is that of small oscillations of a uniform, vertically hanging chain. There the square of the wave speed is proportional to the tension, which increases linearly with distance from the bottom of the chain.

at $z = 0$. From eq. (5.49), we see that this is equivalent to the condition that

$$\frac{d\xi_z}{dz} = 0 \qquad (5.57)$$

at $z = 0$.[†] This gives us the same condition as before, namely $C_2 = 0$.

Because the problem is linear, the other constant C_1 is not defined and merely represents the magnitude of the oscillations. In a real star we would need to set another boundary condition at the centre of the star, and this would determine the possible oscillation frequencies ω. To simulate that here, we take the atmosphere to have a finite depth H at which the base of the atmosphere is fixed.[‡] Thus we demand that $\xi_z = 0$ at $x = H$. If we define τ_1 as being the time for a wave to propagate from the top to the bottom of the atmosphere, i.e.

$$\tau_1 = \int_0^H \frac{dx}{c_s}, \qquad (5.58)$$

then the oscillation frequencies are given by the condition that $\omega\tau_1 = Z_i$, where $Z_i, i = 0, 1, 2, \ldots$, is one of the infinite set of zeros $(Z_0 < Z_1 < Z_2 < \cdots)$ of the Bessel function $J_m(Z)$. We see that the boundary conditions have picked out a discrete set of oscillation frequencies.

5.3 Further reading

The general theory of stellar oscillations can be found in Unno *et al.* (1979) and Cox (1980). Wave propagation in plane-parallel atmospheres is discussed in Lamb (1932, Chap. X). The properties of internal waves in a stratified, incompressible fluid are discussed by Turner (1973, Chap. 2).

5.4 Problems

5.4.1 A constant gravity, convectively neutral, polytropic, plane-parallel atmosphere has a zero density surface at $x = 0$ and a fixed base at $x = H$. Show that the adiabatic vertical velocity perturbations of the atmosphere form a set of eigenfunctions of the form $u_n(x)e^{i\omega_n t}$, which satisfy the following orthogonality relation:

$$\int_0^H f(x)u_n(x)u_m(x)\,dx = \delta_{mn}, \qquad (5.59)$$

where the function $f(x)$ is to be determined.

How do we determine the eigenvalues?

[†] For the problem of the swinging chain, this is simply the condition that the bottom of the chain is free to swing.

[‡] For the chain problem, this is equivalent to taking the top of the chain to be fixed.

5.4.2 A static, gaseous plane-parallel atmosphere rests on a fixed base at $z = 0$ and is subject to a fixed gravitational field $\mathbf{g} = (0, 0, -g(z))$, where $g(z) = \Omega^2 z$ and Ω is a constant. The atmosphere is isothermal, with (isothermal) sound speed c_s. Show that the density structure is of the form $(0 \leq z < \infty)$

$$\rho(z) = \rho_0 \exp(-z^2/2H^2), \tag{5.60}$$

where ρ_0 is the density at $z = 0$ and H is a constant.

The atmosphere is subject to small adiabatic velocity perturbations of the form $\mathbf{u} = (0, 0, w(z)e^{i\omega t})$, where ω is the oscillation frequency. What kind of modes would you expect to find in such an atmosphere, and what kind of modes does this perturbation represent?

Show that $w(z)$ obeys the following equation:

$$\frac{d^2 w}{dz^2} - \frac{z}{H^2}\frac{dw}{dz} + \frac{1}{\gamma}\left(\frac{\omega^2}{c_s^2} - \frac{1}{H^2}\right)w = 0, \tag{5.61}$$

where γ is the usual ratio of specific heats.

By considering a series solution about $z = 0$, or otherwise, show that the oscillation modes $w_n(z)$ with finite kinetic energy $E = \int_0^\infty \frac{1}{2}\rho w^2\, dz$ take the form of polynomials of degree $N = 2n + 1$ for $n = 0, 1, 2, \ldots$. Show that the corresponding oscillation frequencies ω_n are given by

$$\omega_n^2 = \frac{(\gamma N + 1)c_s^2}{H^2}. \tag{5.62}$$

5.4.3 A static polytropic atmosphere subject to uniform gravitational acceleration $\mathbf{g} = g\hat{\mathbf{x}}$ (where x is measured downwards) has density and pressure given by

$$\rho = \rho_0\left(\frac{x}{H}\right)^m, \quad p = p_0\left(\frac{x}{H}\right)^{m+1}, \tag{5.63}$$

below the surface $x = 0$, where ρ_0, p_0, and H are positive constants (in general $1 + 1/m$ differs from the adiabatic exponent $\gamma = 1 + 1/n$ of the gas).

Find the linearized equations for the displacement, density perturbation and pressure perturbation, neglecting self-gravity, and assuming that the displacement has the following form:

$$\text{Re}[\boldsymbol{\xi}(x)\exp(i\mathbf{k}\cdot\mathbf{r} - i\omega t)]. \tag{5.64}$$

Eliminate all variables in terms of the displacement divergence $\Delta = \nabla\cdot\boldsymbol{\xi}$ and obtain the following equation

$$z\frac{d^2\Delta}{dx^2} + (m+2)\frac{d\Delta}{dx} + \left[\frac{n(m+1)}{n+1}\frac{\omega^2}{gk} + \left(\frac{m-n}{n+1}\right)\frac{gk}{\omega^2} - kx\right]k\Delta = 0. \tag{5.65}$$

Make the transformation $\Delta = w(x)e^{-kx}$ and find the solutions of the resulting equation for w as power series in x. Show that the only solutions for which Δ is finite at the surface and decays with depth are those for which w is a polynomial in *x*. Deduce the dispersion relation,

$$n(m+1)\left(\frac{\omega^2}{gk}\right)^2 - (n+1)(2N+m)\left(\frac{\omega^2}{gk}\right) + (m-n) = 0, \qquad (5.66)$$

where N is a positive integer.

Discuss the behaviour of the frequency eigenvalues ω in the limit of large mode number N, and identify these solutions as p- and g-modes. (See Lamb (1932, Chap. X).)

6

Damping and excitation of stellar oscillations

In Chapter 5 we considered the various oscillation modes that a star can show when perturbed in some way. Many types of star are observed to pulsate. Perhaps the best known are the class known as Cepheids (after the prototype star δ Cephei). These pulsate primarily in a radial mode, so that we observe the full amplitude of their oscillations. The period of these oscillations is directly related to the stellar radius, just as the pitch of an organ note is related to the length of the pipe producing it, the lowest notes coming from the longest pipes. Finding the star's radius in this way gives a measure of its brightness, so that simply by comparing the periods of two Cepheids we know their relative brightnesses. Thus, with careful calibration, measuring the period and the apparent brightness of a Cepheid gives its distance. As Cepheids are bright stars they can be seen in very distant galaxies and therefore can be used to produce a distance scale of great importance in astronomy.

However, most stars do not show pulsations of readily observable amplitude. This must mean that for most stars any oscillations set in motion by perturbations received during their lives, or in the process of formation, have long since been damped out in some way. The existence of these damping processes in turn implies that if a star does pulsate, an excitation mechanism for these pulsations must be operating. Typically such mechanisms are not dynamical, such as a stellar collision, but involve physical processes within the star itself. We get considerable insight into what must be involved in the excitation process by plotting the locations of known types of pulsating stars on the Hertzsprung–Russell diagram of luminosity versus effective temperature. Most types of pulsating stars lie in a restricted region (the 'instability strip') of this diagram, that is pulsations are favoured in stars with effective temperatures $T_e \sim 6000$–$10\,000$ K.

In this chapter we take a brief look at damping and excitation processes in stars. As usual, we do so in a very simplified manner in order to bring out the essential physical ideas, without obscuring them with excessive algebraic complexity.

6.1 A simple set of oscillations

To make things as simple as possible, we consider initially one-dimensional acoustic oscillations in a box containing a gas of uniform density ρ and uniform pressure p. Thus we ignore gravity and set $g = 0$. We assume that the perfect gas law holds, so that the internal energy (which is proportional to the temperature, $e \propto T$) is given by the equation of state:

$$p = (\gamma - 1)\rho e. \tag{6.1}$$

We assume that variations only occur in the x-direction. The only non-zero component of the velocity is the x-component, which we write as u. Then the mass conservation equation to linear order is given by

$$\frac{\partial \rho'}{\partial t} + \rho \frac{\partial u}{\partial x} = 0 \tag{6.2}$$

and the momentum equation is given by

$$\rho \frac{\partial u}{\partial t} + \frac{\partial p'}{\partial x} = 0. \tag{6.3}$$

If, as we have assumed before, the perturbations are adiabatic, then the energy conservation equation in the form

$$\frac{De}{Dt} + \frac{p}{\rho} \, \text{div} \, \mathbf{u} = 0 \tag{6.4}$$

linearizes to become

$$\frac{\partial e'}{\partial t} + \frac{p}{\rho} \frac{\partial u}{\partial x} = 0. \tag{6.5}$$

The equation of state in linearized form becomes

$$\frac{p'}{p} = \frac{\rho'}{\rho} + \frac{e'}{e}. \tag{6.6}$$

Since p and ρ are constants, it is now straightforward to combine these equations. We take the time derivative of eq. (6.3), and eliminate p' using eq. (6.6) to yield

$$\frac{\rho}{p} \frac{\partial^2 u}{\partial t^2} = -\frac{\partial}{\partial x} \left[\frac{\partial}{\partial t} \left(\frac{\rho'}{\rho} \right) + \frac{\partial}{\partial t} \left(\frac{e'}{e} \right) \right]. \tag{6.7}$$

Then, using eq. (6.2) to replace ρ', eq. (6.5) to replace e', and also the equation of state, eq. (6.1), we obtain the one-dimensional wave equation:

$$\frac{\partial^2 u}{\partial t^2} = c_s^2 \frac{\partial^2 u}{\partial x^2}, \tag{6.8}$$

where as usual the sound speed is given by $c_s^2 = \gamma p / \rho$.

Again, to keep things simple we assume that the box has fixed ends at $x = 0$ and $x = H$, at which $u = 0$, and we also assume that the modes initially have zero velocity, $u = 0$, at $t = 0$. Then the solutions are given by

$$u_n(x, t) = U \sin(\omega_n t) \sin(n\pi x/H), \tag{6.9}$$

for all integers $n = 1, 2, 3, \ldots$, and where the oscillation frequencies are given by

$$\omega_n = \frac{n\pi c_s}{H}. \tag{6.10}$$

Note that U is an arbitrary velocity amplitude.

For future reference, we note that from eq. (6.5) the perturbed internal energy is given by

$$e_n'(x, t) = \frac{U c_s}{\gamma} \cos(\omega_n t) \cos\left(\frac{n\pi x}{H}\right). \tag{6.11}$$

6.2 Damping by conductivity

We now look at the effect on these oscillation modes of a small thermal conductivity λ. By 'small' in this context, we mean that the effect of the conductivity occurs on a timescale which is much longer than the period $2\pi/\omega_n$ of the oscillation.

Then the only change to the equations we have used above is the addition of a conductivity term in the energy equation. Thus, eq. (6.4) now becomes

$$\frac{De}{Dt} + \frac{p}{\rho} \operatorname{div} \mathbf{u} = \frac{\partial}{\partial x}\left(\lambda \frac{\partial T}{\partial x}\right). \tag{6.12}$$

Using the fact that $e \propto T$, and thus that $e' \propto T'$, this equation then linearizes as follows:

$$\frac{\partial e'}{\partial t} + \frac{p}{\rho} \frac{\partial u}{\partial x} = \Lambda \frac{\partial^2 e'}{\partial x^2}, \tag{6.13}$$

where Λ is a constant which is proportional to the conductivity λ of the unperturbed gas. The analysis then proceeds as before, and the effect of eliminating e' in eq. (6.7) is to add an extra term on the r.h.s. of the wave equation for u (eq. (6.8)), which now takes the following form:

$$\frac{\partial^2 u}{\partial t^2} = c_s^2 \frac{\partial^2 u}{\partial x^2} - (\gamma - 1)\Lambda \frac{\partial^3 e'}{\partial x^3}. \tag{6.14}$$

We seek an approximate solution to this equation, making use of the assumption that Λ is in some sense small. We therefore take the solution to be of the following form:

$$u(x, t) = u_n(x, t) + y(x, t), \tag{6.15}$$

where u_n is given by eq. (6.9) and y is of order Λ, so that $y \ll u_n$. Then, by construction, this solution obeys eq. (6.14) to zeroth order in Λ. To first order in Λ, the equation becomes

$$\frac{\partial^2 y}{\partial t^2} = c_s^2 \frac{\partial^2 y}{\partial x^2} - \frac{(\gamma - 1)\Lambda U c_s}{\gamma} \left(\frac{n\pi}{H}\right)^3 \cos(\omega_n t) \sin\left(\frac{n\pi x}{H}\right). \tag{6.16}$$

We seek a separable solution of this equation in the form

$$y(x, t) = f(t) \sin\left(\frac{n\pi x}{H}\right), \tag{6.17}$$

so that $y = 0$ at $x = 0$ and $x = H$. Here $f(t)$ is an unknown function, which must obey the following equation:

$$\frac{d^2 f}{dt^2} + \omega_n^2 f = -\frac{(\gamma - 1)\Lambda U c_s}{\gamma} \left(\frac{n\pi}{H}\right)^3 \cos(\omega_n t). \tag{6.18}$$

This equation represents an oscillator being resonantly forced at its oscillation frequency. The solution is obtained by looking for a particular integral of the form $f = At \sin(\omega_n t)$ for some constant fA, and is given by

$$u = U \sin(\omega_n t) \sin\left(\frac{n\pi x}{H}\right) \left\{1 - \frac{t}{\tau}\right\}, \tag{6.19}$$

where the (damping) timescale τ is given by

$$\tau = \frac{2\gamma H^2}{(\gamma - 1)n^2 \pi^2 \Lambda}. \tag{6.20}$$

This solution is valid only as long as $y \ll u_n$, or in other words as long as $t \ll \tau$. We can see that the effect of a small conductivity is a slow decrease in the amplitude of the oscillations. In the stellar context, the conduction of heat is carried out by photons, and so this damping of the wave is known as 'radiation damping'.

The physical picture of this process is quite simple (Fig. 6.1). At any instant, the acoustic wave has peaks and troughs in pressure, or equivalently peaks and troughs in temperature. The effect of the conductivity is to try to even out these peaks and troughs, and thus to reduce the amplitude of the wave. Physically the evening-out happens because the parts of the star hotter than the equilibrium temperature cool by emitting more radiation than they absorb, while the reverse happens in parts cooler than equilibrium.

6.2.1 An alternative derivation

It is instructive to look at a different derivation of the effects of damping, and one which more closely mirrors calculations of damping and excitation of oscillations in real stellar models. We start with eq. (6.14), which describes the effect of

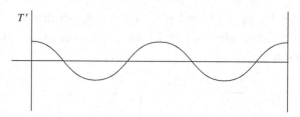

Fig. 6.1. Damping by conductivity. The temperature perturbation T' is plotted at a particular time in the oscillation cycle. Where $T' = 0$, the temperature gradient and the heat flow are greatest. Parts of the star hotter than the equilibrium temperature emit more radiation than they absorb, while the reverse happens in parts cooler than equilibrium.

a conductivity proportional to Λ on the standard wave equation for acoustic modes. We then take the Fourier transform of this equation with respect to time in the form $\exp(\mathrm{i}\omega t)$. To zeroth order in Λ, this equation becomes one for $u(x; \omega)$:

$$c_s^2 \frac{\mathrm{d}^2 u}{\mathrm{d}x^2} + \omega^2 u = 0. \tag{6.21}$$

This is a Sturm–Liouville problem with a set of eigenfunctions $u_n(x)$ and a corresponding set of real eigenvalues ω_n^2. By taking the inverse Fourier transform, we would discover the set of solutions for $u(x, t)$ we found above. We also note that the eigenfunctions are orthogonal, and can be appropriately normalized, so that we may write

$$\int_0^H u_n u_m \, \mathrm{d}x = \delta_{nm}, \tag{6.22}$$

where δ_{nm} is the Kronecker delta.

We now consider the effect of the term in Λ, which involves the third spatial derivative of e'. Because Λ is a small quantity, we can use the Fourier transform of the original energy equation, eq. (6.5), i.e.

$$\mathrm{i}\omega e' = -\frac{p}{\rho} \frac{\mathrm{d}u}{\mathrm{d}x}, \tag{6.23}$$

to eliminate e' from eq. (6.14). Thus we obtain an equation for u in the following form:

$$\frac{(\gamma - 1)\Lambda p}{\mathrm{i}\omega\rho} \frac{\mathrm{d}^4 u}{\mathrm{d}x^4} + c_s^2 \frac{\mathrm{d}^2 u}{\mathrm{d}x^2} + \omega^2 u = 0. \tag{6.24}$$

We now consider the effect of the small conductive term on the eigenfunction u_n, which satisfies the equation

$$c_s^2 \frac{\mathrm{d}^2 u_n}{\mathrm{d}x^2} + \omega_n^2 u_n = 0, \tag{6.25}$$

with eigenfrequency ω_n. To do this we assume as before that the solution to the perturbed equation is of the form $u = u_n + y$, with $y \ll u_n$. But now we can use the fact that the eigenfunctions of a Sturm–Liouville problem form a complete set, so that any function y can be written as an appropriate sum over them. Thus we take

$$u = u_n + \sum_{m \neq n} a_m u_m, \tag{6.26}$$

where the coefficients a_m are of order Λ and are to be determined. In fact for our present purposes we are not interested in finding these, but are, rather, interested in calculating the perturbation to the eigenfrequency ω_n. We therefore substitute eq. (6.26) into eq. (6.24), keeping only terms that are first order in Λ, to yield

$$(\omega_n^2 - \omega^2)u_n + \sum_{m \neq n}\{(\omega_m^2 - \omega^2)a_m u_m\} = \frac{1}{i\omega}\frac{(\gamma - 1)\Lambda p}{\rho}\frac{d^4 u_n}{dx^4}. \tag{6.27}$$

We now multiply both sides of this equation by u_n, integrate from zero to H and use the orthogonal property of the eigenfunctions to give an equation for the new eigenfrequency ω:

$$\omega_n^2 - \omega^2 = \frac{1}{i\omega}\frac{(\gamma - 1)\Lambda p}{\rho}\int_0^H u_n \frac{d^4 u_n}{dx^4}\,dx. \tag{6.28}$$

Since we expect the new eigenfrequency to be close to the original one, we let

$$\omega = \omega_n + i\epsilon, \tag{6.29}$$

where $\epsilon \ll \omega_n$. Then, to first order in ϵ, we find

$$\epsilon = \frac{(\gamma - 1)\Lambda p}{2\omega_n^2 \rho}\int_0^H \left(\frac{d^2 u_n}{dx^2}\right)^2 dx, \tag{6.30}$$

where we have integrated by parts twice. This demonstrates that $\epsilon > 0$. Thus the quantities which formerly oscillated in the form $\exp(i\omega_n t)$ now vary with time as $\exp(i\omega t) = \exp(i\omega_n t - \epsilon t)$. These correspond to oscillations which are damped (since $\epsilon > 0$), with a damping timescale of $\tau = 1/\epsilon$. This solution is now valid for all time $t > 0$ and not just for $0 < t \ll \tau$.

6.3 The effect of heating and cooling – the ϵ-mechanism

For most of their lives, stars radiate because of the release of nuclear energy in their interiors. For a low-mass main-sequence star like the Sun, the energy production process involves the conversion of hydrogen to helium in a set of reactions called the p-p chain, since the first reaction in the process requires the fusion of two protons (or hydrogen nuclei). The energy production rate of this process has the following rough dependence:

$$\epsilon_{nuc}(\rho, T) \propto \rho T^{\alpha}, \tag{6.31}$$

where ϵ_{nuc} is the rate of energy release per unit mass and the index $\alpha \sim 3\text{–}5$. In more massive stars on the main sequence the fusion of hydrogen to helium proceeds by a catalytic reaction cycle involving carbon, nitrogen and oxygen nuclei, called the CNO cycle. The energy production rate for this process can be written as above, but now with $\alpha \sim 16 - 18$. Later in their lives, stars survive by fusing helium nuclei to form carbon. Since it takes three helium nuclei to form a carbon nucleus, this is essentially a three-body process, and so has a stronger dependence on the density. It is also very sensitive to temperature and can typically be written as something like

$$\epsilon_{nuc}(\rho, T) \propto \rho^2 T^{40}. \tag{6.32}$$

Energy generation like this means that the energy equation has a source term, so that

$$\frac{De}{Dt} + \frac{p}{\rho} \operatorname{div} \mathbf{u} = \mathcal{H}(\rho, e), \tag{6.33}$$

where $\mathcal{H} = \rho\epsilon_{nuc}$ is the energy generation rate per unit volume and we have used the fact that for a perfect gas $T \propto e$. In linearized form this equation becomes

$$\frac{\partial e'}{\partial t} + \frac{p}{\rho}\frac{\partial u}{\partial x} = e'\frac{\partial \mathcal{H}}{\partial e}\bigg|_{\rho} + \rho'\frac{\partial \mathcal{H}}{\partial \rho}\bigg|_{e}. \tag{6.34}$$

Proceeding with the analysis as before, the wave equation for u now has some extra terms and becomes

$$\frac{\partial^2 u}{\partial t^2} = c_s^2\frac{\partial^2 u}{\partial x^2} - \left(\frac{1}{e}\frac{\partial \mathcal{H}}{\partial e}\right)\frac{\partial e'}{\partial x} - \left(\frac{1}{e}\frac{\partial \mathcal{H}}{\partial \rho}\right)\frac{\partial \rho'}{\partial x}. \tag{6.35}$$

We saw above that the nuclear energy generation rate in stars typically depends much more strongly on T than on ρ. Thus,

$$\frac{\partial \mathcal{H}}{\partial \ln e} \gg \frac{\partial \mathcal{H}}{\partial \ln \rho}. \tag{6.36}$$

This implies that we can, as a first approximation, neglect the ρ' term in this equation. This also has the beneficial effect of simplifying the algebra. Including this term adds little of interest and serves mainly to complicate the analysis. So we use the equation in the following form:

$$\frac{\partial^2 u}{\partial t^2} = c_s^2\frac{\partial^2 u}{\partial x^2} - \frac{1}{e^2}\frac{\partial \mathcal{H}}{\partial \ln e}\frac{\partial e'}{\partial x}. \tag{6.37}$$

To study stability we again assume that the heating term has only a small effect over an individual oscillation cycle, so that we can treat this term as small. Thus we start again with a particular oscillation mode which would be present in the absence of the heating term. As before we choose

$$u_n(x, t) = U \sin(\omega_n t) \sin(n\pi x / H), \tag{6.38}$$

for which

$$\frac{\partial e'}{\partial x} = -\frac{U c_s}{\gamma} \frac{n\pi}{H} \cos(\omega_n t) \sin(n\pi x / H). \tag{6.39}$$

Then we assume that adding the small heating term leads, in a first approximation, to a solution of the following form:

$$u(x, t) = u_n(x, t) + y(x, t), \tag{6.40}$$

where $y \ll u_n$. In this case the equation for y is given by

$$\frac{\partial^2 y}{\partial t^2} = c_s^2 \frac{\partial^2 y}{\partial x^2} + \frac{1}{e^2} \frac{\partial \mathcal{H}}{\partial \ln e} \frac{U c_s}{\gamma} \frac{n\pi}{H} \cos(\omega_n t) \sin\left(\frac{n\pi x}{H}\right). \tag{6.41}$$

This is very similar to eq. (6.16), we had before, and by essentially the same analysis we deduce that the oscillations grow in amplitude if $\partial \mathcal{H} / \partial \ln e > 0$. This makes physical sense. During the oscillation cycle, when the gas is most compressed, it is also hottest. If $\partial \mathcal{H} / \partial \ln e > 0$, the energy production mechanism adds yet more heat to the gas at this stage and the oscillation grows. This is known as overstability.

This effect is no longer thought to be the driving mechanism for most stellar pulsations. What happens instead is that either the damping effects of conductivity overwhelm this type of excitation, or the temperature sensitivity drives the nuclear burning regions of the star convective. On a historical note, the possibility of this mechanism led Sir James Jeans (1929) to the conclusion that nuclear fusion was unlikely to be the process which powers stars.

6.4 The effect of opacity – the κ-mechanism

The mechanism which is now thought to drive most observed large-scale stellar oscillations is one which taps the outward heat flux. It is caused by the temperature sensitivity of the thermal conductivity (or, in stellar parlance, the opacity κ). To investigate this mechanism we need an equilibrium state with a temperature gradient and a non-zero heat flux down this gradient. For this reason we use as our illustrative base state the polytropic atmosphere, whose oscillation properties we discussed in Chapter 4.

There we used the coordinate x to measure distance from the top of the atmosphere at $x = 0$, and we assumed a solid base at $x = H$. The pressure and density were related by $p \propto \rho^{1+1/m}$, and we found $\rho \propto x^m$, $p \propto x^{m+1}$ and $T \propto e \propto x$.

Perturbing this atmosphere, assuming purely vertical motions, the linearized mass conservation equation is given by

$$\frac{\partial \rho'}{\partial t} + u\frac{\partial \rho}{\partial x} + \rho\frac{\partial u}{\partial x} = 0 \tag{6.42}$$

and the linearized equation of motion is given by

$$\rho\frac{\partial u}{\partial t} = -\frac{\partial p'}{\partial x} + \rho' g, \tag{6.43}$$

where the gravity $g > 0$ is fixed and given by hydrostatic equilibrium:

$$g = -\frac{1}{\rho}\frac{dp}{dx}. \tag{6.44}$$

We assume that the atmosphere is a perfect gas, so that $p = (\gamma - 1)\rho e$, and the perturbed equation of state yields

$$\frac{p'}{p} = \frac{\rho'}{\rho} + \frac{e'}{e}. \tag{6.45}$$

Using the mass conservation equation to substitute for div $\mathbf{u}$, the thermal equation, eq. (6.12), becomes

$$\rho\frac{De}{Dt} = \frac{p}{\rho}\frac{D\rho}{Dt} + \rho\,\text{div}(\lambda\nabla T). \tag{6.46}$$

Using the equation of state, this can then be written as follows:

$$\frac{D\rho}{Dt} = \frac{1}{c_s^2}\frac{Dp}{Dt} - \frac{\rho}{\gamma e}\,\text{div}(\lambda\nabla T). \tag{6.47}$$

In the equilibrium situation, this equation implies that

$$\lambda\frac{dT}{dx} = \text{const.} \tag{6.48}$$

Since $dT/dx = $ constant, this implies a form for the dependence of the conductivity λ on density and temperature such that it is constant through the atmosphere. Since $\rho \propto T^m$, we can write this dependence as $\lambda(\rho, T) = F(\rho/T^m)$ for some arbitrary function F. (Note that this is in general not true for the real physical conductivity, but is forced on us by our simplifying assumption that the atmosphere is polytropic. This does not affect the essential physics of our treatment of the κ-mechanism.)

To keep the algebra simple, we now further assume that the unperturbed atmosphere is neutrally stable (or marginally unstable) to convection, i.e. $\gamma = 1 + 1/m$. This means that $A = 0$, or equivalently that

$$\nabla\rho = \frac{1}{c_s^2}\nabla p. \tag{6.49}$$

In this case the linearized version of the thermal equation, eq. (6.46), becomes

$$\frac{\partial \rho'}{\partial t} = \frac{1}{c_s^2}\frac{\partial p'}{\partial t} - Q',$$
(6.50)

where Q' is the Eulerian perturbation of the quantity Q representing the heat flux:

$$Q = \frac{\rho}{\gamma e}\mathrm{div}(\lambda \nabla T).$$
(6.51)

6.4.1 The underlying oscillations: $Q' = 0$

We start by considering the oscillations which occur when there is no perturbation to the heat flux, i.e. $Q' = 0$. In fact, we worked out the detailed properties of the oscillations in this case in Chapter 4. Here we do not need the details, but need to derive some of their properties, making use of the fact that this is a Sturm–Liouville problem.

When $Q' = 0$ and the atmosphere is convectively neutral, the thermal equation is simply given by

$$p' = c_s^2 \rho'.$$
(6.52)

Using this, eqs. (6.42) and (6.43) can be combined to yield

$$\frac{\partial^2 u}{\partial t^2} = \frac{\partial}{\partial x}\left\{\frac{c_s^2}{\rho}\frac{\partial}{\partial x}(\rho u)\right\}.$$
(6.53)

We now Fourier analyse in time in the form $u \propto \exp(i\omega t)$ and obtain the eigenvalue equation describing the oscillation modes as follows:

$$\frac{d}{dx}\left\{\frac{c_s^2}{\rho}\frac{d}{dx}(\rho u_n)\right\} + \omega_n^2 u_n = 0.$$
(6.54)

The oscillation modes are orthogonal, and can be normalized, so that

$$\int_0^H \rho u_m u_n \, dx = \delta_{mn}$$
(6.55)

(see Problem 5.4.1).

6.4.2 Perturbing the underlying oscillations: $Q' \neq 0$

We now allow the heat flux to vary so that $Q' \neq 0$. At the same time we assume that the effect of this term is small in the sense that it takes many oscillation periods to have an effect. Thus it will produce small perturbations in the oscillation frequency ω_n and in the eigenfunction $u_n(x)$.

With this extra term, eq. (6.53) becomes

$$\frac{\partial^2 u}{\partial t^2} = \frac{\partial}{\partial x}\left\{\frac{c_s^2}{\rho}\frac{\partial}{\partial x}(\rho u)\right\} + \frac{1}{\rho}\frac{\partial}{\partial x}[c_s^2 Q'].\tag{6.56}$$

We now consider the perturbation to the mode $u_0 = u_n(x)\sin(\omega_n t)$ in the following form:

$$u(x,t) = u_n(x)\sin(\omega_n t) + y(x,t),\tag{6.57}$$

where y is small. Substituting this in eq. (6.56), and using the fact that u_n satisfies eq. (6.53), we find that y satisfies the following equation:

$$\frac{\partial^2 y}{\partial t^2} - \frac{\partial}{\partial x}\left[\frac{c_s^2}{\rho}\frac{\partial}{\partial x}(\rho y)\right] = G(x)\sin(\omega_n t).\tag{6.58}$$

Here $G(x)$ is some function of x which comes from substituting the solution $u_n(x)$, and the corresponding density $\rho_n(x)$ and temperature $T_n(x)$, into the term Q'.

As before we look for a solution of the form $y = u_n(x)f(t)$, which yields

$$\left[\frac{d^2 f}{dt^2} - \omega_n^2 f\right]u_n(x) = G(x)\sin(\omega_n t).\tag{6.59}$$

We now use the orthogonality properties of the eigenfunctions. We multiply this equation by ρu_n and integrate from 0 to H. This gives an equation for the function $f(t)$ as follows:

$$\frac{d^2 f}{dt^2} - \omega_n^2 f = \left[\int_0^H \rho u_n(x)G(x)\,dx\right]\sin(\omega_n t).\tag{6.60}$$

This is a resonantly forced harmonic oscillator, and, as we found before, whether or not the oscillations grow or damp depends on the sign of the term in square brackets. This sign depends on the functional form of the dependence of the thermal conductivity λ on the density ρ and temperature T. Thus if we choose a form of λ which makes the square bracket positive, the perturbations will grow and the star oscillates.

The physics of the instability for this choice of λ is not easy to see in simple terms. What happens in essence is that the temperature and density dependence of the conductivity manipulate the heat flux through the atmosphere in such a way that there is a net heat gain by the oscillating gas when it is compressed (higher density and so higher temperature), raising its temperature still further. In a real star the temperature sensitivity of the stellar opacity needed to make this mechanism work occurs at the point where hydrogen changes from being predominantly neutral (low opacity) to predominantly ionized (high opacity). Then the hottest gas experiences a sudden increase in opacity as it gains a little heat, so absorbing more of the stellar radiation flux and raising the temperature still further.

Such hydrogen ionization zones require temperatures $\sim 6000 - 10\,000$ K. For this effect to drive oscillations, there must not be too much mass above the ionization zone, so the surface (effective) temperature T_e must have values of this order also. This explains the position of the main instability strip in the Hertzsprung–Russell diagram. For stars with little hydrogen there is another instability strip at higher effective temperatures corresponding to helium ionization.

6.5 Further reading

More details on excitation mechanisms for stellar oscillations, as well as the computation of excitation and damping timescales, can be found in Unno *et al.* (1979) and in Cox (1980). On a historical note, readers might be intrigued by the discussion of Jeans (1929, Chap. IV), who uses a simple stability analysis to try to constrain possible stellar energy-generation mechanisms.

7

Magnetic instability in a static atmosphere

In Chapter 4 we considered the stability of a static fluid configuration against convective instabilities, or buoyancy. We found that for stability the specific entropy must increase upwards. In this chapter we again consider the stability of a static atmosphere, but with the complication of an added magnetic field. We shall find that the magnetic field can act either to stabilize or to de-stabilize the fluid. It is possible to derive a variational principle for perturbations of a fluid containing a magnetic field, just as we did for a non-magnetic fluid in Chapter 4 (see Problem 4.7.4). Of course, the expressions we would derive in doing so contain all the information required to decide stability. But we found in Chapter 4 that we had to manipulate carefully the expressions we derived in the variational principle to extract a useful stability criterion – the Schwarzschild criterion. Adding a magnetic field makes the expressions in the variational principle much more complex, simply because the geometry of the magnetic field and its interaction with the fluid add more degrees of freedom, and there is no simple stability criterion. Accordingly we adopt a simpler, less comprehensive, approach here.

There are some guiding concepts with which a theoretical astrophysicist should be familiar, and we illustrate these here. We discuss the two distinct, but often confused, modes of instability – the buoyancy instability and the Parker instability. As before we keep try to keep the situation simple, in order to bring out the physics of the situation without obscuring it in mathematical detail. Even so, the inclusion of a magnetic field to the perturbation equations does considerably increase the degree of mathematical complexity.

7.1 Magnetic buoyancy

We consider an atmosphere with uniform gravity $\mathbf{g} = (0, 0, -g)$ in Cartesian coordinates and with a horizontal magnetic field $\mathbf{B} = (0, B(z), 0)$. The equation of hydrostatic equilibrium is simply given by

$$\frac{d}{dz}\left(p + \frac{1}{2}B^2\right) = -\rho g. \tag{7.1}$$

We consider an equilibrium magnetic field which is zero in the upper half-space, i.e. $\mathbf{B} = 0$ for $z > 0$, and uniform in the lower half-space, $\mathbf{B} = (0, B_0, 0)$ for $z < 0$, where B_0 is a constant. Thus the field has a sharp boundary at $z = 0$, and there is a sheet current of the form $\mathbf{j} = (J\delta(z), 0, 0)$ flowing in the $z = 0$ plane.

Integrating eq. (7.1) across the $z = 0$ plane, we see that the total pressure P (gas plus magnetic, i.e. $P = p + B_0^2/2$) must be continuous across $z = 0$. Thus if the gas pressure just below $z = 0$, i.e. at $z = 0^-$, is $p = p_0$, the gas pressure just above, at $z = 0^+$, must be $p = p_0 + \Delta p$, where $\Delta p = B_0^2/2$. We assume that the gas obeys the perfect gas law, $p = (\mathcal{R}/\mu)\rho T$. We must also require that the temperature T is continuous at $z = 0$, for otherwise there would be an infinite heat flux at that point, however small the conductivity. We conclude from all this that there must be a jump in the density at $z = 0$. If the density at $z = 0^-$ is $\rho = \rho_0$ and the density at $z = 0^+$ is $\rho = \rho_1$, then

$$\frac{\mathcal{R}\rho_1 T}{\mu} = \frac{\mathcal{R}\rho_0 T}{\mu} + \frac{1}{2}B^2. \tag{7.2}$$

Alternatively we may write $\rho_1 = \rho_0 + \Delta\rho$, where

$$\frac{\Delta\rho}{\rho_0} = \frac{V_A^2/2}{\mathcal{R}T/\mu}. \tag{7.3}$$

Here, V_A is the Alfvén speed in $z < 0$, where $V_A^2 = B_0^2/\rho_0$.

Thus equilibrium requires that we have a heavy fluid above a light one in a gravitational field. This happens because the magnetic field provides pressure, but no mass. It seems likely that this situation is unstable, and, as we now show, this is indeed the case. To simplify the analysis still further, we consider instability over a region very close to the plane $z = 0$, in particular a region which is much smaller than the atmospheric scaleheight $H \sim p_0/g\rho_0$. This means that we can assume the unperturbed pressures and densities to be uniform. Moreover we shall consider only perturbations for which the perturbed velocity obeys div $\mathbf{u} = 0$. This is a useful device which implies that Lagrangian density perturbations vanish, $\delta\rho = 0$, and so cuts out the possibility of acoustic waves. Since we expect any instability in this situation to be caused by gravity (g-modes) rather than by sound waves (p-modes), this has the effect of simplifying the analysis without excluding any of the essential physics.

We consider a perturbation to the magnetic field of the form $\mathbf{B} = \mathbf{B}_0 + \mathbf{b}$, where $\mathbf{b} = (b_x, b_y, b_z)$ is small, and a small resulting velocity of the form $\mathbf{u} = (u, v, w)$.

The linearized induction equation is then given by

$$\frac{\partial \mathbf{b}}{\partial t} = \text{curl}(\mathbf{u} \wedge \mathbf{B}_0). \tag{7.4}$$

Expanding this expression and using the imposed condition div $\mathbf{u} = 0$, we obtain

$$\frac{\partial \mathbf{b}}{\partial t} = B_0 \frac{\partial \mathbf{u}}{\partial y}. \tag{7.5}$$

The linearized form of the momentum equation, including the magnetic terms, is given by

$$\rho \frac{\partial \mathbf{u}}{\partial t} = -\nabla p' - \mathbf{B}_0 \wedge (\text{curl } \mathbf{b}). \tag{7.6}$$

In terms of components, this expands to yield

$$\rho \frac{\partial u}{\partial t} = -\frac{\partial p'}{\partial x} + B_0 \left[\frac{\partial b_x}{\partial y} - \frac{\partial b_y}{\partial x} \right], \tag{7.7}$$

$$\rho \frac{\partial v}{\partial t} = -\frac{\partial p'}{\partial y} \tag{7.8}$$

and

$$\rho \frac{\partial w}{\partial t} = -\frac{\partial p'}{\partial z} + B_0 \left[\frac{\partial b_z}{\partial y} - \frac{\partial b_y}{\partial z} \right]. \tag{7.9}$$

We note first that by taking the curl of eq. (7.6), and using eq. (7.5), we obtain an equation for the vorticity, $\boldsymbol{\omega} = \nabla \wedge \mathbf{u}$, of the perturbed velocity field:

$$\frac{\partial^2 \boldsymbol{\omega}}{\partial t^2} = V_A^2 \frac{\partial^2 \boldsymbol{\omega}}{\partial y^2}. \tag{7.10}$$

This implies that a perturbation which has vorticity is propagated away along the field lines as an Alfvén wave in the y-direction. Such perturbations therefore do not lead to instability. Thus we need only consider perturbations which have zero vorticity, for which curl $\mathbf{u} = 0$. For such perturbations we may write

$$\mathbf{u} = -\nabla \Psi, \tag{7.11}$$

where Ψ is the velocity potential, and, since div $\mathbf{u} = 0$, we also have that Ψ obeys Laplace's equation:

$$\nabla^2 \Psi = 0. \tag{7.12}$$

We now take Fourier transforms so that all perturbed quantities vary as $\propto \exp \{i\omega t + i\mathbf{k} \cdot \mathbf{r}\}$. To ensure that Ψ, and therefore the other quantities, obey Laplace's equation, we need to place a constraint on the components of $\mathbf{k}$.

In the lower half-space $z < 0$, we need to write

$$\Psi = \Psi_0 \exp \{i\omega t + ik_x x + ik_y y + (k_x^2 + k_y^2)^{1/2} z\}, \qquad (7.13)$$

where Ψ_0 is a constant, the magnitude of $k_z = k_\perp = (k_x^2 + k_y^2)^{1/2}$ is chosen such that Laplace's equation is satisfied, and its sign is chosen to ensure that the perturbation vanishes at large distances, i.e. $\Psi \to 0$ as $z \to -\infty$.

Similarly in the upper half-space $z > 0$, we need to write

$$\Psi = \Psi_1 \exp \{i\omega t + ik_x x + ik_y y - (k_x^2 + k_y^2)^{1/2} z\}, \qquad (7.14)$$

where Ψ_1 is a constant, the magnitude of $k_z = k_\perp$ is chosen such that Laplace's equation is satisfied, and its sign is chosen to ensure that the perturbation vanishes at large distances, i.e. $\Psi \to 0$ as $z \to +\infty$.

There are two boundary conditions which we must apply at $z = 0$. First we must ensure that the vertical velocity is continuous across the boundary. Otherwise the two fluid regions would come apart! The vertical component of the velocity is given by $w = \partial \Psi / \partial z$. Evaluating these in the upper and lower half-planes, and equating them at $z = 0$, we obtain

$$\Psi_0 = -\Psi_1. \qquad (7.15)$$

Second, by integrating eq. (7.9) across the boundary, $\int_{-\epsilon}^{\epsilon} dz$, and then letting $\epsilon \to 0$, we find that the perturbation to the total pressure $P = p + B^2/2$ (gas plus magnetic) given by

$$P' = p' + B_0 b_y \qquad (7.16)$$

must be continuous across the boundary.

We note that the boundary in the perturbed fluid is at

$$z = \zeta \exp \{i\omega t + ik_x x + ik_y y\}, \qquad (7.17)$$

where ζ is the z-component of the Lagrangian displacement and is related to the vertical velocity perturbation by $w = i\omega\zeta$. Thus, just above the boundary at $z = 0^+$ we have

$$i\omega\zeta = +k_\perp \Psi_1, \qquad (7.18)$$

and just below the boundary at $z = 0^-$ we have

$$i\omega\zeta = -k_\perp \Psi_0. \qquad (7.19)$$

Since the boundary is no longer at $z = 0$ we must apply the condition that it is the Lagrangian perturbation of the total pressure, $\delta P = P' + \boldsymbol{\xi} \cdot \nabla P = P' - \zeta\rho g$, which is continuous at $z = 0$. Note that here we have used the equilibrium condition $\nabla P = (0, 0, -g\rho)$. Thus to apply the second boundary condition we have that

$$p_1' - \zeta\rho_1 g = p_0' + B_0 b_y - \zeta\rho_0 g, \qquad (7.20)$$

where both sides are to be evaluated at $z = 0$. To do so we need expressions for p_1', p_0' and b_y. From the y-component of eq. (7.5), using the fact that $v = -\partial \Psi / \partial y$, we have in $z < 0$ that

$$i\omega b_y = B_0 k_y^2 \Psi_0. \tag{7.21}$$

From eq. (7.8) we have in $z < 0$ that

$$i\omega \Psi_0 = \frac{p_0'}{\rho_0} \tag{7.22}$$

and in $z > 0$ that

$$i\omega \Psi_1 = \frac{p_1'}{\rho_1}. \tag{7.23}$$

Putting all this together, the second boundary condition (eq. (7.20)) becomes

$$i\omega \rho_1 \Psi_1 - \rho_1 g \frac{k_\perp \Psi_1}{i\omega} = i\omega \rho_0 \Psi_0 + B_0^2 \frac{k_y^2 \Psi_0}{i\omega} + \rho_0 g \frac{k_\perp \Psi_0}{i\omega}. \tag{7.24}$$

Tidying this up, and using the first boundary condition (which yielded $\Psi_0 = -\Psi_1$), this gives a dispersion relation for the modes:

$$\omega^2 = -\frac{g k_\perp (\rho_1 - \rho_0) - \rho_0 V_A^2 k_y^2}{\rho_0 + \rho_1}. \tag{7.25}$$

This fluid configuration is unstable for those modes with $\omega^2 < 0$, i.e. for those modes with

$$k_y^2 < \frac{g k_\perp}{V_A^2} \frac{\Delta \rho}{\rho_0}. \tag{7.26}$$

The most unstable modes, i.e. those with the most negative value of ω^2, are those with $k_y = 0$. These modes have no y-dependence. Since the field is in the y-direction, this implies that the fluid motions are all perpendicular to the field. Thus the field lines are not stretched, and so no extra magnetic energy is created. From an energy point of view these modes can tap gravitational energy by interchanging fluid elements of different density in the gravitational field, without paying a penalty in magnetic energy by having to stretch field lines.

This instability is simple magnetic buoyancy . It is the analogue of the convective instability and comes about because the magnetic field provides the pressure but no mass. Thus in pressure equilibrium matter containing magnetic field is lighter than matter without. As we have seen, it operates even when the fluid is incompressible.

7.2 The Parker instability

The concept of a buoyancy instability driven by magnetic fields contributing pressure but no mass is easily visualized when the fluids (or the perturbations) are incompressible. However, most astronomical fluids are not incompressible, and the modes by which buoyancy can drive an instability are different from those derived above. We sketch here the analysis of a simple problem given by Parker (1979), which serves to illustrate the physical differences.

We consider again a stratified atmosphere with fixed gravity, $\mathbf{g} = (0, 0, -g)$, and horizontal magnetic field, $\mathbf{B}_0 = (0, B_0(z), 0)$, lying in the y-direction. Thus, although $\operatorname{div} \mathbf{B}_0 = 0$, we have $\operatorname{curl} \mathbf{B}_0 \neq 0$, and there is a current proportional to $-\partial B_0 / \partial z$ in the x-direction. We take the equation of state of the unperturbed atmosphere to be isothermal, so that density and pressure are related by $p = c_{\mathrm{is}}^2 \rho$, where c_{is} is the *isothermal* sound speed. For simplicity we assume that the magnetic field is such that the magnetic pressure is a constant fraction α of the gas pressure, i.e. $B_0^2 / 2 = \alpha p$. Then the equilibrium equation (eq. (7.1)) yields the following solutions:

$$p(z) = p_0 \exp(-z/H), \tag{7.27}$$

$$\rho(z) = \rho_0 \exp(-z/H) \tag{7.28}$$

and

$$B_0(z) = B_{00} \exp(-z/2H), \tag{7.29}$$

where the scaleheight H is given by

$$H = \frac{(1 + \alpha) c_{\mathrm{is}}^2}{g}. \tag{7.30}$$

Here p_0, ρ_0 and B_{00} are all constants, being values of the quantities at the reference level $z = 0$.

We now write down the linearized perturbation equations. The equation of mass conservation is given by

$$\frac{\partial \rho'}{\partial t} + w \frac{d\rho}{dz} = -\rho \operatorname{div} \mathbf{u}. \tag{7.31}$$

We assume that the perturbations are adiabatic, so that $\delta p = (\gamma p / \rho) \delta \rho$. Since the basic atmosphere is isothermal, this means, according to the Schwarzschild criterion, that, in the absence of a magnetic field, it is stable to convection. The mass conservation equation becomes

$$\frac{\partial p'}{\partial t} = \gamma c_{\mathrm{is}}^2 \frac{\partial \rho'}{\partial t} - \frac{(\gamma - 1) c_{\mathrm{is}}^2 \rho w}{H}. \tag{7.32}$$

The linearized induction equation,

$$\frac{\partial \mathbf{b}}{\partial t} = \operatorname{curl}(\mathbf{u} \wedge \mathbf{B}_0), \tag{7.33}$$

has three components:

$$\frac{\partial b_x}{\partial t} = B_0 \frac{\partial u}{\partial y}, \tag{7.34}$$

$$\frac{\partial b_y}{\partial t} = -B_0 \frac{\partial u}{\partial x} - B_0 \frac{\partial w}{\partial z} + B_0 \frac{w}{2H} \tag{7.35}$$

and

$$\frac{\partial b_z}{\partial t} = B_0 \frac{\partial w}{\partial y}. \tag{7.36}$$

In the linearized momentum equation we need to recall that curl $\mathbf{B}_0 \neq 0$, so it is given by

$$\rho \frac{\partial \mathbf{u}}{\partial t} = -\nabla p' - \mathbf{B}_0 \wedge (\text{curl } \mathbf{b}) - \mathbf{b} \wedge (\text{curl } \mathbf{B}_0) + \rho' \mathbf{g}. \tag{7.37}$$

This has the following three components:

$$\rho \frac{\partial u}{\partial t} = -\frac{\partial p'}{\partial x} + B_0 \left\{ \frac{\partial b_x}{\partial y} - \frac{\partial b_y}{\partial x} \right\}, \tag{7.38}$$

$$\rho \frac{\partial v}{\partial t} = -\frac{\partial p'}{\partial y} - B_0 \left\{ \frac{b_z}{2H} \right\} \tag{7.39}$$

and

$$\rho \frac{\partial w}{\partial t} = -\frac{\partial p'}{\partial z} + B_0 \left\{ \frac{\partial b_z}{\partial y} - \frac{\partial b_y}{\partial z} + \frac{b_y}{2H} \right\} - g\rho'. \tag{7.40}$$

In total we now have eight partial differential equations for the eight variables $\mathbf{b}, \mathbf{u}, p'$ and ρ'. If the unperturbed atmosphere were uniform, the obvious thing to do next would be to Fourier transform with all variables $\propto \exp(i\omega t + i\mathbf{k} \cdot \mathbf{r})$. We would then be left with eight linear, homogeneous, algebraic equations, leaving an 8×8 determinant to be evaluated to give us the dispersion relation. Parker's trick was to realize that something similar can be achieved by making the following assumptions about the behaviour of the variables. We take

$$p' \propto \exp(i\omega t + i\mathbf{k} \cdot \mathbf{r}) \times \exp(-z/2H), \tag{7.41}$$

$$\rho' \propto \exp(i\omega t + i\mathbf{k} \cdot \mathbf{r}) \times \exp(-z/2H), \tag{7.42}$$

$$\mathbf{u} \propto \exp(i\omega t + i\mathbf{k} \cdot \mathbf{r}) \times \exp(+z/2H) \tag{7.43}$$

and

$$\mathbf{b} \propto \exp(i\omega t + i\mathbf{k} \cdot \mathbf{r}). \tag{7.44}$$

These relationships imply that both the magnetic energy perturbation, b^2, and the kinetic energy perturbation, ρu^2, are independent of z. By making these substitutions we then find that the resulting algebraic equations are linear and homogeneous. Importantly the exponential z-dependence of the quantities describing the unperturbed atmosphere cancels out. We are then left with an 8×8 determinant to evaluate in order to obtain the dispersion relation. This necessitates a large amount of messy, but straightforward, algebra, which we leave to the reader's imagination and omit. To make the algebra simpler, the resulting dispersion relation is given by Parker (1979) in terms of dimensionless quantities. We define a dimensionless frequency Ω in terms of the time for an isothermal wave to cross a scaleheight,

$$\Omega = \frac{\omega H}{c_{is}}, \qquad (7.45)$$

and a dimensionless wavevector, in terms of the scaleheight,

$$\mathbf{q} = H\mathbf{k}. \qquad (7.46)$$

With these the full dispersion relation is given by

$$\Omega^4 - \Omega^2(2\alpha + \gamma)[q_y^2 + q_z^2 + 1/4] + q_y^2\{2\alpha\gamma(q_y^2 + q_z^2 + 1/4) - (1+\alpha)(1+\alpha-\gamma)\}$$

$$+ \frac{q_x^2}{2\alpha(q_x^2 + q_y^2) - \Omega^2}[\gamma\Omega^4 - \Omega^2\{2\alpha\gamma q_y^2 - 2\alpha(2\alpha + \gamma)q_z^2 + (\gamma-1) + (1/2)\alpha\gamma\}$$

$$- 4\alpha^2\gamma q_y^2(q_z^2 + 1/4)] = 0. \qquad (7.47)$$

This is a sixth-order equation for Ω, or because of time-symmetry a cubic equation for Ω^2. On physical grounds, we expect the roots to come in pairs, with one pair representing (magnetically modified) acoustic waves, one representing (magnetically modified) buoyancy waves and one representing magnetic (torsional or Alfvén) waves. In the previous section we were able to simplify the analysis by removing the acoustic waves by setting div $\mathbf{u} = 0$ and the magnetic waves by taking curl $\mathbf{u} = 0$. Here we do not have that luxury. Nevertheless, we still expect any instability to come through buoyancy waves. However, we have already seen that in the absence of a field the unperturbed atmosphere is buoyantly stable.

We do not attempt a general analysis of the dispersion relation, but rather consider two simple types of modes.

7.2.1 Modes with $k_y = 0$

Since the unperturbed magnetic field lies in the y-direction, if we set $k_y = 0$, or equivalently $q_y = 0$ or $\partial/\partial y = 0$, the fluid motions are perpendicular to the field and so do not stretch the field lines. In the incompressible case, these were the most unstable modes, because such fluid motions did not change the magnetic energy. The dispersion relation now becomes

$$\Omega^2[\Omega^4 - \Omega^2(2\alpha + \gamma)(q_x^2 + q_z^2 + 1/4) + q_x^2(\alpha(\alpha + \gamma) + \gamma - 1)] = 0. \quad (7.48)$$

Now two of the modes have zero frequency, and so are neutrally stable. This is because the perturbation in this case does not bend or twist the magnetic field lines , so that there are no magnetic waves. The remaining equation takes the following form:

$$\Omega^4 - \mathcal{B}\Omega^2 + \mathcal{C} = 0, \tag{7.49}$$

where $\mathcal{B} > 0$, implying that the sum of the roots is positive. Thus there is a negative root for Ω^2, and therefore instability, if and only if $\mathcal{C} < 0$. Thus instability occurs if and only if

$$\gamma < 1 - \alpha. \tag{7.50}$$

Since the unperturbed atmosphere is isothermal, we already know that if there is no field ($\alpha = 0$) then it is unstable to convection if and only if $\gamma < 1$, which does not occur for physical fluids. If a field is added ($\alpha > 0$) we require an even smaller value of γ to achieve instability. This means that adding a magnetic field for these modes (which are the most unstable modes in the incompressible case) stabilizes them. This happens because for fluid motions perpendicular to the field, the equation describing the dragging of field lines by the fluid becomes

$$\frac{D}{Dt}\left(\frac{B}{\rho}\right) = 0. \tag{7.51}$$

Thus, as the fluid moves the local field varies as $B \propto \rho$. This means that the magnetic pressure, p_M, varies as $p_M = B^2/2 \propto \rho^2$. Thus $p_M \propto \rho^2$, and the magnetic field acts like a gas with $\gamma = 2$. In an isothermal atmosphere this is stabilizing.

7.2.2 Modes with $k_x = 0$, $k_y \neq 0$

We now look at modes which vary along the field lines, but where all field lines have no x-dependence and so remain locally parallel to each other; thus, the field lines are distorted. In the incompressible case we found that this contributed a stabilizing influence. The dispersion relation now becomes

$$\Omega^4 - \Omega^2(2\alpha + \gamma)(q_y^2 + q_z^2 + 1/4) - q_y^2\{(1+\alpha)(1+\alpha-\gamma)$$
$$- 2\alpha\gamma(q_y^2 + q_z^2 + 1/4)\} = 0. \tag{7.52}$$

We have removed the factor of Ω^2 which corresponds to the two fast magnetosonic modes. These are not excited because the mode does not compress the field lines. Once again this equation is of the following form:

$$\Omega^4 - \mathcal{B}\Omega^2 + \mathcal{C} = 0, \tag{7.53}$$

with $\mathcal{B} > 0$. Thus we again have instability if and only if $\mathcal{C} < 0$. This requires that

$$q_y^2 + q_z^2 < \frac{(1+\alpha)(1+\alpha-\gamma)}{2\alpha\gamma} - \frac{1}{4}. \tag{7.54}$$

It is then straightforward to show that there is a non-vanishing range of unstable modes with $q_y^2 + q_z^2 > 0$ if and only if

$$\gamma < \frac{(1+\alpha)^2}{1+3\alpha/2}. \tag{7.55}$$

Since the r.h.s. is a monotonically increasing function of α for $\alpha > 0$, this implies that, for any value of γ, there is instability for some large enough value of α, i.e. for a large enough field strength.

We now need to ask ourselves why these modes, which were stable for an incompressible fluid (the limit $\gamma \to \infty$), become unstable once compressibility is permitted. The simple answer is that compressibility permits movement of the fluid along the field lines. These modes produce undulating field lines. The fluid can then fall to the troughs of the undulations and become denser there, while fluid at the peaks of the undulations becomes less dense. These motions release gravitational energy. If the gravitational energy released is more than the magnetic energy required to undulate the field lines, instability occurs.

7.3 Further reading

More detailed discussion of the analysis given here can be found in Parker (1979, Chap. 13). Further discussion, and the derivation of a variational principle, can be found in Chandrasekhar (1961).

7.4 Problems

7.4.1 A perfectly conducting fluid with density $\rho(z)$ and pressure $p(z)$ lies in a constant vertical gravitational field g and is permeated by a constant vertical magnetic field $\mathbf{B}_0 = (0, 0, B)$. It experiences a velocity perturbation of the following form:

$$\mathbf{u} = [u(z), v(z), w(z)] \exp\{i(k_x x + k_y y + \omega t)\}, \tag{7.56}$$

and the perturbed magnetic field is $\mathbf{B} = \mathbf{B}_0 + \mathbf{b}$. Explain physically why we might wish to consider perturbations for which $\mathrm{div}\,\mathbf{u} = 0$.

For such perturbations show that

$$i\omega\mathbf{b} = B\frac{d\mathbf{u}}{dz}. \tag{7.57}$$

Show that the perturbation equations can be written as fourth-order differential equations for the vertical component of the velocity perturbation, w, in the following form:

$$\frac{d}{dz}\left(\rho\frac{dw}{dz}\right) + \frac{B^2}{\omega^2}\left(\frac{d^2}{dz^2} - k^2\right)\frac{d^2w}{dz^2} = k^2\rho w + \frac{gk^2}{\omega^2}\left(\frac{d\rho}{dz}\right)w, \qquad (7.58)$$

where $k^2 = k_x^2 + k_y^2$.

Using suitable boundary conditions at fixed boundaries z_1 and z_2, show that ω^2 is real.

What waves do these perturbations represent? Under what physical conditions might we expect ω^2 to be negative? (See Chandrasekhar (1961, Chap. X).)

8

Thermal instabilities

In this chapter we consider instabilities generated by the effects of heating and cooling, coupled with the effect of thermal conductivity . The traditional application of these ideas is to the interstellar medium, but the ideas we generate here have wider applicability. Thermal instability is extremely common in astronomy, where we often deal with luminous objects attempting to lose large amounts of heat energy.

We consider a fluid which is initially uniform and fills all space. Since we are interested in thermal effects, we neglect the effects of gravity. Then the equations of interest are mass conservation,

$$\frac{\partial \rho}{\partial t} + \mathrm{div}(\rho \mathbf{u}) = 0, \tag{8.1}$$

momentum conservation,

$$\rho \frac{\mathbf{D u}}{\mathbf{D} t} + \nabla p = 0, \tag{8.2}$$

and thermal energy conservation in the form

$$\frac{1}{\gamma - 1} \frac{\mathrm{D} p}{\mathrm{D} t} - \frac{\gamma}{\gamma - 1} \frac{p}{\rho} \frac{\mathrm{D} \rho}{\mathrm{D} t} = \mathrm{div}(\lambda \nabla T) + Q(\rho, T), \tag{8.3}$$

where $Q(\rho, T)$ is the net heat gain per unit volume per unit time. We write $Q = Q^+ - Q^-$, where Q^+ is the heating rate and Q^- is the cooling rate.

Cooling of astrophysical gases occurs when electrons and molecules in excited states decay to states of lower energy, emitting photons as they do so. The electrons may be bound to atoms or free, and excitation may occur by collisions with other particles or by radiation. For a fixed mass in a fixed volume, the resulting cooling rate, Q^-, is fixed by the excitation rates, and is thus simply a function of temperature, T. In Fig. 8.1 we show a schematic description of $Q^-(T)$ for the interstellar medium. A striking feature of this plot is that Q^- is a multi-valued function of T.

Thermal instabilities

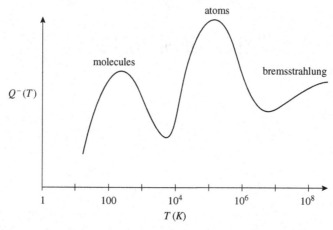

Fig. 8.1. Typical radiative cooling curve for the interstellar medium. The two peaks correspond to cooling by molecules and atoms, respectively. At higher temperatures the gas is almost completely ionized, and cooling is mainly by transitions between free electron states as thermal electrons are deflected by ions (free–free, or thermal bremsstrahlung, emission).

Typical heating processes in the interstellar medium include heating by cosmic rays and by ultraviolet radiation from hot stars. Cosmic rays are energetic elementary particles, and they give up their energy to a target fluid by colliding with its constituent particles. These processes are effectively independent of the thermodynamic state of the fluid. Hence the cosmic ray heating rate for a fixed mass of fluid is independent of temperature. Comparing this with the multi-valued radiative cooling function of the interstellar medium, we see that, for a gas in which thermal equilibrium balances radiative cooling against cosmic ray heating, there will be typically more than one possible equilibrium temperature.

In this chapter we consider the stability of these equilibrium points. If, as often happens, there is more than one possible temperature at which stable equilibrium can occur, we shall ask what happens at the interface between two such regions at different temperatures.

8.1 Linear perturbations and the Field criterion

To give a specific example we consider a simple paradigm in which the cooling function $Q^-(T)$ looks like a cubic, and the heating function $Q^+(T)$ is a constant, as shown in Fig. 8.2(a). The three intersections of these two curves give the temperatures at which thermal equilibrium can occur. The net cooling function $Q = Q^+ - Q^-$ is sketched in Fig. 8.2, and the equilibrium temperatures are given by $Q = 0$.

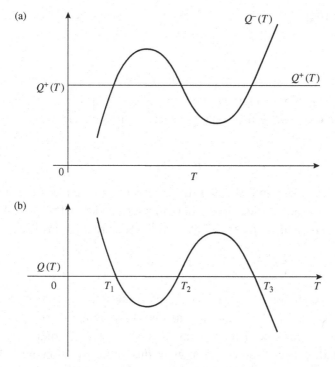

Fig. 8.2. (a) Schematic cubic cooling function Q^- for interstellar medium balanced by a constant heating function Q^+. (b) Resulting net heating function $Q = Q^+ - Q^-$. Thermal equilibrium is possible at the three temperatures T_1, T_2 and T_3, where $Q = 0$.

We start with a uniform gas at rest. We then linearize the equations and assume that all quantities vary as $\propto \exp\{i\omega t + i\mathbf{k} \cdot \mathbf{x}\}$. Then the mass conservation equation becomes

$$i\omega\rho' + \rho i\mathbf{k} \cdot \mathbf{u} = 0, \tag{8.4}$$

the momentum equation becomes

$$i\omega\rho\mathbf{u} + i\mathbf{k}p' = 0 \tag{8.5}$$

and the energy equation becomes

$$\frac{i\omega}{\gamma - 1}p' - \frac{i\omega\gamma p}{(\gamma - 1)\rho}\rho' = -\lambda k^2 T' + Q'. \tag{8.6}$$

Here Q' is the perturbed net heating rate, which we can write as follows:

$$Q' = Q_\rho\rho' + Q_T T', \tag{8.7}$$

where $Q_\rho = (\partial Q/\partial \rho)_T$ and $Q_T = (\partial Q/\partial T)_\rho$. We also need the linearized equation of state, which for a perfect gas is given by

$$\frac{p'}{p} = \frac{\rho'}{\rho} + \frac{T'}{T}. \tag{8.8}$$

We note from eq. (8.5) that **k** is parallel to **u**, so that any motions will be compressive. We eliminate **u** by taking the scalar product of eq. (8.5) with **k** and substituting in eq. (8.4). This yields

$$p' = \frac{\omega^2}{k^2}\rho'. \tag{8.9}$$

Now eqs. (8.6), (8.7) and (8.9) are three linear homogeneous equations for the quantities ρ', p' and T'. Setting the determinant of the coefficients of these three quantities to zero, a little algebra gives the dispersion relation in the following form:

$$\frac{i\omega}{\gamma - 1}\left(\frac{\omega^2}{k^2} - c_s^2\right) - (Q_T - \lambda k^2)\frac{T}{p}\left(\frac{\omega^2}{k^2} - \frac{c_s^2}{\gamma}\right) = Q_\rho. \tag{8.10}$$

This is a cubic equation for ω. The equations of motion give us two factors of ω, as we found before in our derivation for the dispersion relation for sound waves, and the extra factor of ω comes from the time derivative in the thermal energy equation. Full and detailed discussion of the roots of this cubic equation are to be found in Field (1965). Here we note the implications in a few simple cases.

8.1.1 Acoustic waves

We note first that if we eliminate all thermal effects by setting $\lambda = 0$ and $Q_\rho = Q_T = 0$, the dispersion relation reduces to

$$\omega^2 = k^2 c_s^2, \tag{8.11}$$

which (as expected) we recognise as the relation describing acoustic waves, previously derived in Chapter 2.

8.1.2 No net heating or cooling, but small conductivity

If we now set $Q = 0$ and allow a small, non-zero, conductivity λ, then we can show (see Problem 8.4.1) that the dispersion relation takes the following form:

$$\omega = \pm k c_s + i\delta, \tag{8.12}$$

where $\delta \propto \lambda$ and $\delta > 0$, independent of the sign of $\pm k$. Thus, whichever direction the sound wave moves, the amplitude of the wave is damped exponentially. In terrestrial situations where there is little geometrical attentuation, this is the main way in which sound waves are damped.

8.1.3 Slow cooling

In many astrophysical situations, and in particular in the interstellar medium, we are interested in the effects of heating and cooling which operate on much longer timescales than the sound crossing timescale for a particular lengthscale. This means that we are interested in the root of eq. (8.10), which is such that $\omega \ll kc_s$. In this limit, which corresponds to the pressure being constant, eq. (8.10) becomes

$$i\omega = \frac{(\gamma - 1)T}{\gamma p}(Q_T - \lambda k^2) - \frac{(\gamma - 1)\rho}{\gamma p}Q_\rho. \tag{8.13}$$

If the conductivity is negligible, we may set $\lambda = 0$, and, using the fact that $p \propto \rho T$, the equation becomes

$$i\omega = \frac{\gamma - 1}{\gamma p} \left.\frac{\partial Q}{\partial \ln T}\right|_p. \tag{8.14}$$

We conclude that

$$\text{instability} \Leftrightarrow \left.\frac{\partial Q}{\partial \ln T}\right|_p > 0. \tag{8.15}$$

This is known as the Field stability criterion.

This makes physical sense. Suppose we are in equilibrium at some temperature T, so that $Q = 0$, and that the inequality is satisfied. Then because net heating or cooling occurs so slowly, evolution occurs at fixed pressure. Now suppose that the temperature is increased slightly to $T + \delta T$, with $\delta T > 0$. Then, according to the inequality, the net heating Q becomes positive and the temperature continues to climb. Similarly, if $\delta T < 0$, the net heating becomes negative and the temperature continues to drop.

If there is a non-zero conductivity ($\lambda \neq 0$), the equation becomes

$$i\omega = \frac{\gamma - 1}{\gamma p} \left.\frac{\partial Q}{\partial \ln T}\right|_p - \lambda k^2 \frac{(\gamma - 1)T}{\gamma p}. \tag{8.16}$$

This implies that if k is large enough, $i\omega$ becomes negative and the equilibrium becomes stable. Thus sufficiently small lengthscales (corresponding to large k) are stable, because on such small lengthscales the conductivity acts quickly enough to smooth out any temperature fluctuations. The critical lengthscale at which this happens is the Field length, $\lambda_F = 2\pi/k_F$, found by setting $\omega = 0$ in eq. (8.16). Hence,

$$k_F^2 = \left(\frac{2\pi}{\lambda_F}\right)^2 = \frac{1}{\lambda T} \left.\frac{\partial Q}{\partial \ln T}\right|_p. \tag{8.17}$$

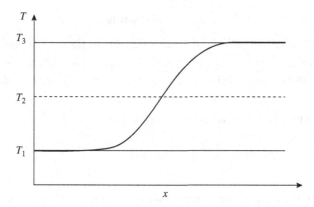

Fig. 8.3. Temperature profile at the interface between cool and hot phases of the interstellar medium.

8.2 Heating and cooling fronts

Suppose now that the net heating function $Q(T)$ at fixed pressure, p, is schematically of the form shown in Fig. 8.2. This has three possible equilibrium temperatures, $T_1 < T_2 < T_3$, of which, by the Field criterion, only T_1 and T_3 are stable. Thus some of the gas can be stable at a low temperature T_1 and some at a high temperature T_3. This means that there can be interfaces between these two temperature phases. In this section we consider what happens at such interfaces.

At such an interface the temperature profile $T(x)$ as a function of position x might look something like that sketched in Fig. 8.3. The gas at temperatures $T_1 < T < T_2$ undergoes net cooling and the gas at temperatures $T_2 < T < T_3$ undergoes net heating. Thus these heating and cooling processes try to steepen the temperature gradient at the interface. At the same time, the conductivity acts in the opposite direction, trying to reduce the temperature gradient. It is possible that some kind of global balance is achieved and the interface stays fixed with some equilibrium shape. In general, however, for a given temperature distribution $T(x)$, the net heating of gas at temperatures $T_2 < T < T_3$ does not balance the net cooling of gas at temperatures $T_1 < T < T_2$. This means that, integrated over the interface as a whole, there is either a net heating or a net cooling of the gas. If there is a net heating, this implies that the position of the interface moves in the direction of the cooler gas (the hot gas manages to incorporate some cool gas). Conversely, if there is a net cooling, the position of the interface moves in the direction of the hotter gas (the cool gas incorporates some hot gas).

We now look at the equations governing the motions of such hot or cold fronts in one spatial dimension x. In line with the assumptions behind the Field criterion, we assume that any motions associated with the movement of the interface are highly subsonic. Then we may neglect the velocity terms in the momentum equation in one dimension so that it becomes

$$\frac{\partial p}{\partial x} = 0, \tag{8.18}$$

or equivalently $p = \text{const.}$

In one dimension it is sensible to change to a Lagrangian variable $m(x, t)$ defined as follows:

$$m(x, t) = \int_{-\infty}^{x} \rho(x', t)\, \mathrm{d}x'. \tag{8.19}$$

Using the mass conservation equation, we can then show that

$$\left.\frac{\partial m}{\partial t}\right|_x = -\rho u, \tag{8.20}$$

where u is the velocity in the x-direction. Hence we deduce that

$$\left.\frac{\partial}{\partial x}\right|_t = \rho \left.\frac{\partial}{\partial m}\right|_t \tag{8.21}$$

and that

$$\left.\frac{\partial}{\partial t}\right|_x = \left.\frac{\partial}{\partial t}\right|_m - \rho u \left.\frac{\partial}{\partial m}\right|_t. \tag{8.22}$$

These show that the Lagrangian derivative, Df/Dt, for any quantity f becomes

$$\left.\frac{\partial f}{\partial t}\right|_x + u \left.\frac{\partial f}{\partial x}\right|_t = \left.\frac{\partial f}{\partial t}\right|_m. \tag{8.23}$$

Using all these, the one-dimensional energy equation in the form

$$\frac{1}{\gamma - 1}\frac{Dp}{Dt} - \frac{\gamma}{\gamma - 1}\frac{p}{\rho}\frac{D\rho}{Dt} = \frac{\partial}{\partial x}\left(\lambda \frac{\partial T}{\partial x}\right) + Q \tag{8.24}$$

becomes

$$-\frac{\gamma}{\gamma - 1}\frac{p}{\rho}\left.\frac{\partial \rho}{\partial t}\right|_m = \rho \frac{\partial}{\partial m}\left(\lambda \rho \frac{\partial T}{\partial m}\right) + Q. \tag{8.25}$$

We now use the equation of state, $p = (\mathcal{R}/\mu)\rho T$, together with the subsonic approximation $p = \text{const.}$, to eliminate $\rho \propto 1/T$. Thus we can regard both λ and Q as functions of T alone. Then by scaling the time variable as follows:

$$\tau = \frac{\gamma - 1}{\gamma}\frac{p}{(\mathcal{R}/\mu)^2} t \tag{8.26}$$

and writing

$$\mathcal{L}(T) = \left(\frac{\mathcal{R}/\mu}{p}\right)^2 \frac{Q(T)}{T}, \tag{8.27}$$

the equation governing the evolution of the temperature front becomes

$$\frac{\partial T}{\partial t} = \frac{\partial}{\partial m}\left(\frac{\lambda(T)}{T}\frac{\partial T}{\partial m}\right) + \mathcal{L}(T). \tag{8.28}$$

A useful approximation is often to take $\lambda(T) \propto T^\alpha$ for some α (for example, conductivity by thermal electrons has $\alpha = 5/2$ and conductivity by neutral ions has $\alpha = 1/2$). In this case, with suitable scaling, the energy equation becomes

$$\frac{\partial T}{\partial t} = \frac{\partial}{\partial m}\left(T^{\alpha-1}\frac{\partial T}{\partial m}\right) + \mathcal{L}(T). \tag{8.29}$$

For a steadily moving front, with velocity U in m-space, we can set $\xi = m - Ut$ and obtain an ordinary differential equation for T, regarding U as an eigenvalue. With the substitution $T = Z^{1/\alpha}$ we have

$$Z^\beta\frac{d^2Z}{d\xi^2} + U\frac{dZ}{d\xi} + F(Z) = 0, \tag{8.30}$$

where $\beta = (\alpha - 1)/\alpha$ and $F(Z) = \alpha T^{\alpha\beta}\mathcal{L}(T)$.

8.3 Further reading

A full description of the analysis presented here on the thermal stability of the interstellar medium is given by Field (1965). A more general discussion and application of the evolution of heating and cooling fronts is given by Meerson (1996). Also of interest are the papers by Elphick, Regev & Spiegel (1991) and Elphick, Regev & Shaviv (1992).

8.4 Problems

8.4.1 Show that the dispersion relation for linear waves propagating in a uniform compressible gas with constant thermal conductivity λ is given by

$$i\omega(\omega^2 - k^2c_s^2) + \lambda(\gamma - 1)(k^2T/p)[\omega^2 - k^2c_s^2/\gamma] = 0. \tag{8.31}$$

Show that the effect of a small thermal conductivity is always to damp travelling sound waves.

8.4.2 For a fluid with conductivity $\lambda \propto T$, show that with suitable scalings a one-dimensional heating/cooling front satisfies the following equation:

$$\frac{\partial T}{\partial t} = \frac{\partial^2 T}{\partial m^2} + \mathcal{L}(T), \tag{8.32}$$

where m is the Lagrangian variable.

If $\mathcal{L} = T(1 - T^2)$, show that there is a solution $T(m)$ to this equation for which the front is stationary. Sketch the solution and give physical reasons why the front is stationary in this case.

Show that a general steady solution of the equation extremises the functional

$$\mathcal{F}[T] = \int \left[\frac{1}{2} \left(\frac{\partial T}{\partial m} \right)^2 + U(T) \right] dm, \tag{8.33}$$

where we define $\mathcal{L}(T) = -\partial U / \partial T$, and that

$$\frac{\partial T}{\partial t} = -\frac{\delta \mathcal{F}}{\delta T}, \tag{8.34}$$

where $\delta/\delta T$ is the functional derivative.

Show also that

$$\frac{d\mathcal{F}}{dt} = -\int \left(\frac{\delta \mathcal{F}}{\delta T} \right)^2 dm \tag{8.35}$$

and deduce that $\mathcal{F}$ is a Liapunov functional.

Denoting the steady solution of the equation by $T^*(m)$, show that

$$\frac{d^2 T^*}{dm^2} = -\frac{d}{dt^*}[V(T^*)], \tag{8.36}$$

where we define $V = -U$.

Note that eq. (8.36) is analogous to the equation of motion of a particle of unit mass moving in a potential V. Hence show that if the net heating function is of the form $\mathcal{L}(T) = T(1 - T^2) + A$, where A is a constant, there is a steady front only if $A = 0$. (See Elphick *et al.* (1991).)

8.4.3 The interstellar medium is modelled as a perfect gas, with equation of state $p = \mathcal{R}\rho T / \mu$, subject to cooling per unit volume at the rate $\epsilon(\rho, T) = -\rho^2 \Lambda(T)$ and with thermal conductivity $\lambda(T) = \lambda_0 T^\alpha$, where λ_0 is a constant and $\alpha > 0$. Gravity is neglected. Explain briefly the circumstances under which it is reasonable to assume that the pressure remains uniform, i.e. $\nabla p = 0$.

In this case, show that a planar one-dimensional flow obeys the following equation:

$$\frac{1}{\gamma - 1} \frac{\partial p}{\partial t} + \frac{\gamma}{\gamma - 1} p \frac{\partial v}{\partial x} + \rho^2 \Lambda - \frac{\partial}{\partial x} \left(\lambda \frac{\partial T}{\partial x} \right) = 0, \tag{8.37}$$

where v is the velocity in the x-direction.

Show further that if the flow remains at constant pressure then

$$\frac{\partial T}{\partial t} + v \frac{\partial T}{\partial x} + \left(\frac{\gamma - 1}{\gamma} \right) \left(\frac{\mu}{\mathcal{R}} \right)^2 p \frac{\Lambda(T)}{T}$$
$$- \left(\frac{\gamma - 1}{\gamma} \right) \frac{\lambda_0 T}{p} \frac{\partial}{\partial x} \left(T^\alpha \frac{\partial T}{\partial x} \right) = 0.$$

Using the Lagrangian variable

$$m(x, t) = \int_0^x \rho(x, t) \, dx \tag{8.38}$$

and an appropriately scaled time $\tau = Ct$, where the constant C is to be determined, show that this equation can be written in the following form

$$\frac{\partial T}{\partial \tau} + \frac{\Lambda(T)}{T} = \lambda_0 \frac{\partial}{\partial m} \left(T^{\alpha - 1} \frac{\partial T}{\partial m} \right) = 0. \tag{8.39}$$

At time $t = 0$, gas fills the half-space $x > 0$ and has uniform temperature $T = T_0$. The region $x < 0$ contains cold ($T = 0$) infinitely dense gas which does not move but cools infinitely fast. The gas in $x > 0$ cools only by thermal conduction (i.e. $\Lambda = 0$ if $T > 0$). Explain why it is reasonable to seek a similarity solution of the form

$$T(m, \tau) = T_0 f(\xi), \tag{8.40}$$

with similarity variable $\xi = m/(\lambda T_0^{\alpha-1}\tau)^{1/2}$. Write down appropriate boundary conditions for $f(\xi)$ at $\xi = 0$ and as $\xi \to \infty$.

In the case when $\lambda(T) = \lambda_0 T$, find the function $f(\xi)$ in terms of the function $\text{erf}(z) = (2/\sqrt{\pi}) \int_0^z \exp(-s^2)\, ds$, defined so that $\text{erf}(\infty) = 1$.

Sketch the resulting solution $T(m, \tau)$, indicating the behaviour as τ increases.

Show that the rate L at which energy is radiated by the gas at $x \leq 0$ varies as $L \propto t^{-1/2}$.

9

Gravitational instability

In this chapter we consider instability driven by the self-gravity of the fluid. This is clearly an extremely important process whenever we deal with the formation of bound objects such as stars and planets, and more generally large-scale structure such as galaxies and clusters.

9.1 The Jeans instability

To start with a simple picture we consider first a fluid at rest, with uniform density ρ and uniform pressure p, filling the whole of space. Then if we perturb the fluid, the linearized equation of motion is given by

$$\frac{\partial \mathbf{u}}{\partial t} = -\frac{\nabla p'}{\rho} - \nabla \Phi', \tag{9.1}$$

where $\mathbf{u}$ is the fluid velocity, p' is the Eulerian pressure perturbation and Φ' is the Eulerian perturbation of the gravitational potential Φ. The linearized mass conservation equation is given by

$$\frac{\partial \rho'}{\partial t} + \rho \operatorname{div} \mathbf{u} = 0, \tag{9.2}$$

where ρ' is the Eulerian density perturbation. The linearized version of Poisson's equation, relating gravitational potential to mass density, is given by

$$\nabla^2 \Phi' = 4\pi G \rho'. \tag{9.3}$$

Assuming that the perturbations are adiabatic, the density and pressure perturbations are related through the following equation:

$$p' = c_{\mathrm{s}}^2 \rho', \tag{9.4}$$

where the sound speed c_{s} is given by $c_{\mathrm{s}}^2 = \gamma p / \rho$.

Since the initial state is both uniform and at rest, we can Fourier analyze these equations in both time and space, or equivalently we can take all linear quantities to have space and time dependence of the form $\exp\{i(\omega t + \mathbf{k} \cdot \mathbf{r})\}$.

Then, using eq. (9.4), eq. (9.1) becomes

$$i\omega\mathbf{u} = -i\mathbf{k}c_{\mathrm{s}}^2\frac{\rho'}{\rho} - i\mathbf{k}\Phi', \tag{9.5}$$

eq. (9.2) becomes

$$i\omega\frac{\rho'}{\rho} + i\mathbf{k} \cdot \mathbf{u} = 0 \tag{9.6}$$

and eq. (9.3) becomes

$$-k^2\Phi' = 4\pi G\rho'. \tag{9.7}$$

We now eliminate $\mathbf{u}$ by taking the scalar product of eq. (9.5) with $\mathbf{k}$ and substituting into eq. (9.6), and then we use eq. (9.7) to eliminate Φ'. This gives a single linear homogeneous equation for ρ'/ρ, whose coefficient then gives us the dispersion relation:

$$\omega^2 = k^2 c_{\mathrm{s}}^2 - 4\pi G\rho. \tag{9.8}$$

In the absence of gravity, i.e. setting $G = 0$, we recognize this relation as giving simple acoustic waves, with sound speed c_{s}. These are then modified by the presence of gravity, which acts with a frequency $\omega_{\mathrm{G}} = (4\pi G\rho)^{1/2}$, or equivalently acts on a timescale $\tau_{\mathrm{G}} = 2\pi/\omega_{\mathrm{G}} = (\pi/G\rho)^{1/2}$. If the gravity term becomes large enough, which occurs at sufficiently large wavelengths that $k^2 c_{\mathrm{s}}^2 < 4\pi G\rho$, we see that $\omega^2 < 0$ and instability sets in. The critical wavelength at which this occurs, that is the value of $2\pi/k$ at which $\omega^2 = 0$, is known as the Jeans length λ_{J} given by

$$\lambda_{\mathrm{J}}^2 = \frac{\pi c_{\mathrm{s}}^2}{G\rho}. \tag{9.9}$$

We can also define a critical mass, called the Jeans mass, $M_{\mathrm{J}} = \rho\lambda_{\mathrm{J}}^3$. This yields

$$M_{\mathrm{J}} = \left(\frac{\pi}{G}\right)^{3/2}\frac{c_{\mathrm{s}}^3}{\rho^{1/2}}. \tag{9.10}$$

There is a simple physical interpretation of this result. Consider an element of the fluid of size $\sim\lambda$. Gravity acts on this fluid element to try to make it collapse on a timescale τ_{G}. In the other direction, the fluid element can use pressure to support itself against gravity. But to do so the two sides of the fluid element need to be in (pressure) communication with each other. Since pressure communication occurs at the speed of sound, the timescale on which the two sides can communicate is just the sound crossing time for the element, i.e. $\tau_{\mathrm{s}} = \lambda/c_{\mathrm{s}}$. We then see from the above that $\tau_{\mathrm{G}} = \tau_{\mathrm{s}}$ when $\lambda = \lambda_{\mathrm{J}}$. Thus for elements of size $\lambda > \lambda_{\mathrm{J}}$ pressure communication cannot occur quickly enough to prevent gravitational collapse.

Although the result we obtain from this analysis is a simple one and affords a straightforward physical explanation, the derivation of the result is flawed. The reason for this is that the initial, unperturbed state was not in equilibrium, in the sense that it did not satisfy the unperturbed equations. The equations it did not satisfy were the unperturbed Poisson equation,

$$\nabla^2 \Phi = 4\pi G \rho, \tag{9.11}$$

together with the assumption that $\nabla \Phi = 0$ in the unperturbed state. If ρ is uniform and fills all space, then these conditions cannot be satisfied.

One tempting way out of this dilemma is to imagine that the results here might apply to the very centre of a large mass of gas, where we might be able to take approximately $p \approx$ const., $\rho \approx$ const. and $\nabla \Phi \approx 0$. Unfortunately, the timescale on which the instability operates is exactly the timescale ($\sim 1/\sqrt{G\rho}$) on which such a cloud would collapse due to gravity. Thus no matter how large the cloud, from the point of view of looking at an instability, we cannot assume that the cloud is at rest. Problem 9.5.3 shows that perturbations of such collapsing flows grow algebraically rather than exponentially.

All is not lost, however. If we consider the fluid lying at the bottom of a large shallow and *fixed* gravitational potential well caused by something else (for example dark matter) then the fluid there does obey $p \approx$ const., $\rho \approx$ const. and $\nabla \Phi \approx 0$. Then in the region of the bottom of the well, the above analysis does apply approximately to perturbations of short enough wavelength.

9.2 Isothermal, self-gravitating plane layer

At the present time in the Universe, star formation occurs mainly in the discs of spiral galaxies. Thus it is important to consider the gravitational stability of a plane layer of gas. We consider first the equilibrium configuration.

9.2.1 Equilibrium configuration

We consider a layer of gas at rest and centred on the plane $z = 0$. The gas is uniform in the x- and y-directions and extends to infinity in both these directions. We take the gas to be isothermal, so that the equation of state is simply $p = \rho c_s^2$, where $c_s^2 =$ const. Then hydrostatic equilibrium in the z-direction implies

$$\frac{1}{\rho}\frac{dp}{dz} + \frac{d\Phi}{dz} = 0. \tag{9.12}$$

We now define a variable m as follows:

$$m = \int_0^z \rho \, dz, \tag{9.13}$$

which is simply the surface density of material between 0 and z. Then, also using the equation of state, the equation of hydrostatic equilibrium becomes

$$c_s^2 \frac{d\rho}{dm} = -\frac{d\Phi}{dz}.$$ (9.14)

If we assume that the gravitational potential also depends only on z, Poisson's equation is given by

$$\frac{d^2\Phi}{dz^2} = 4\pi G\rho.$$ (9.15)

Combining these equations we eliminate Φ and obtain an equation for ρ as follows:

$$c_s^2 \frac{d^2\rho}{dm^2} = -4\pi G.$$ (9.16)

We now impose the condition that ρ is symmetric above and below the $z = 0$ (or $m = 0$) plane and define the total surface density of the (half) layer as M, so that

$$M = \int_0^\infty \rho \, dz.$$ (9.17)

Then the solution to eq. (9.16) is as follows:

$$\rho(m) = \frac{2\pi GM^2}{c_s^2} \left[1 - \left(\frac{m}{M}\right)^2 \right].$$ (9.18)

In order to obtain the density structure as a function of physical height z, it is convenient to define a quantity ξ

$$\xi = \frac{m}{M}.$$ (9.19)

Then eq. (9.18) is simply given by

$$\rho(\xi) = \rho_0(1 - \xi^2),$$ (9.20)

where ρ_0 is the density on the midplane $z = 0$. Then since by definition $\rho = dm/dz = M \, d\xi/dz$, we see that

$$M \frac{d\xi}{dz} = \rho_0(1 - \xi^2),$$ (9.21)

which integrates as follows:

$$\rho_0 z = M \tanh^{-1}(\xi).$$ (9.22)

We substitute for ξ in eq. (9.20) and, using the standard identity $1 - \tanh^2 x = \mathrm{sech}^2 x$, we find the equilibrium density structure of the layer as follows:

$$\rho(z) = \rho_0 \mathrm{sech}^2(z/H),$$ (9.23)

where the scaleheight H is given by

$$H = \frac{M}{2\rho_0},\tag{9.24}$$

or, using the definition of ρ_0,

$$H = \frac{c_s^2}{4\pi GM}.\tag{9.25}$$

We note that we have again side-stepped the issue of whether the equilibrium configuration can physically exist. The solution for the gravitational potential $\Phi(z)$ does not tend to a constant for large $|z|$. Instead, $\Phi \rightarrow |z|$ as $z \rightarrow \pm\infty$. So, once again, we cannot provide a fully self-consistent picture.

9.2.2 Stability analysis

We now consider small oscillatory perturbations to the self-gravitating layer of the form $\propto \exp\{i(\omega t)\}$. The linearized equation of motion is given by

$$-\omega^2 \boldsymbol{\xi} = -\frac{1}{\rho}\nabla p' + \frac{\rho'}{\rho^2}\nabla p - \nabla\Phi'.\tag{9.26}$$

To avoid the complications of considering oscillations arising from buoyancy, we assume that the perturbations, like the equilibrium structure, are isothermal. This implies that

$$p' = c_s^2 \rho',\tag{9.27}$$

where $c_s^2 = \text{const.}$ Using this, the equation of motion is given by

$$-\omega^2 \boldsymbol{\xi} = -\nabla W - \nabla\Phi',\tag{9.28}$$

where we have defined

$$W = \frac{p'}{\rho}.\tag{9.29}$$

Rather than looking for the general solution for the oscillation modes (stable and unstable) of the layer, we are interested in the critical point at which the layer becomes unstable to self-gravity. From our earlier analysis (Chapter 4), we know from the 'exchange of stabilities' that as we pass from stable to unstable configurations, the quantity ω^2 must pass through zero. Thus to find the critical configuration on the border between stability and instability we can simply set $\omega^2 = 0$. On doing this, we find that eq. (9.28) becomes

$$\Phi' = -W.\tag{9.30}$$

We now substitute this into the linearized version of Poisson's equation:

$$\nabla^2 \Phi' = 4\pi G\rho'\tag{9.31}$$

and use the definition of W to obtain

$$\nabla^2 W + \frac{4\pi G \rho}{c_s^2} W = 0. \tag{9.32}$$

We consider modes of a particular wavelength $\lambda = 2\pi/k$ in the horizontal direction by Fourier analyzing in x and y, that is by writing $W = W(z)\exp\{i(k_x x + k_y y)\}$. The equation then becomes

$$\frac{d^2 W}{dz^2} + \left(\frac{4\pi G \rho(z)}{c_s^2} - k^2\right) W = 0, \tag{9.33}$$

where $k^2 = k_x^2 + k_y^2$.

In solving this equation we have to take account of the structure of the layer in the z-direction. To do so we replace the independent variable z by ξ using eqs. (9.20) and (9.22). After a little algebra, we obtain the equation for $W(\xi)$ as follows:

$$\frac{d^2 W}{d\xi^2} + \frac{2\xi}{1-\xi^2}\frac{dW}{d\xi} + \left[\frac{2}{1-\xi^2} - \frac{kM/\rho_0}{(1-\xi^2)^2}\right] W = 0. \tag{9.34}$$

We now recall that the solutions of Legendre's equation,

$$\frac{d^2 y}{dx^2} + \frac{2x}{1-x^2}\frac{dy}{dx} + \left[\frac{\nu(\nu+1)}{1-x^2} - \frac{\mu^2}{(1-x^2)^2}\right] W = 0, \tag{9.35}$$

which are non-singular at $x = \pm 1$, are called associated Legendre polynomials denoted by $P_\nu^\mu(x)$. Their regularity requires that ν be a positive integer and that μ be an integer in the range $-\nu \le \mu \le \nu$.

Thus in the equation for $W(\xi)$ we see that $\nu = 1$, and for non-zero k we require $\mu = 1$. This gives the critical wavenumber $k_J = \rho_0/M$ at which instability sets in. Using the definition of ρ_0, we can write the critical wavenumber as follows:

$$k_J = \frac{2\pi GM}{c_s^2}. \tag{9.36}$$

From our physical picture in Section 9.1 we expect that modes with $k < k_J$ are unstable to self-gravity. In terms of the scaleheight of the layer, we see that modes with wavelengths $\lambda > \lambda_J$ are unstable, where the Jeans length λ_J is given by

$$\lambda_J = 4\pi H. \tag{9.37}$$

9.3 Stability of a thin slab

Seen from distances much larger than the scaleheight H, the layer discussed above appears very thin and the internal structure of the layer becomes irrelevant. Moreover, we have found that the critical scalelength on which gravitational instability sets in is many times H. Thus it also seems likely that the internal structure of the layer does not play a critical role in determining the Jeans length λ_J. For this reason, we now consider the stability of an infinitesimally thin slab of fluid situated in the plane $z = 0$. Moreover, we are interested in motions due to self-gravity, and we expect these to be only in the plane of the slab.

As an equilibrium configuration we consider the slab to have constant surface density $\Sigma(x, y) = \Sigma_0 = $ const. and constant pressure $P(x, y) = P_0 = $ const. Then the equilibrium gravitational potential is given by

$$\Phi_0 = -2\pi G \Sigma_0 |z|. \tag{9.38}$$

Now we consider perturbations of the equilibrium of the form $\propto \exp\{i(\omega t + k_x x + k_y y)\}$. The mass conservation equation is given by

$$\frac{\partial \Sigma}{\partial t} + \text{div}(\Sigma \mathbf{u}) = 0, \tag{9.39}$$

where we recall that the velocity $\mathbf{u}$ is only in the (x, y)-plane. The linearized version of this is then given by

$$i\omega \Sigma' + i\mathbf{k} \cdot \mathbf{u} \Sigma_0 = 0. \tag{9.40}$$

Similarly, the equation of motion given by

$$\frac{\partial \mathbf{u}}{\partial t} + \mathbf{u} \cdot \nabla \mathbf{u} = -\frac{1}{\Sigma} \nabla_\perp P - \nabla_\perp \Phi, \tag{9.41}$$

where the operator $\nabla_\perp$ acts only in the (x, y)-plane, linearizes to yield

$$i\omega \mathbf{u} = -\frac{1}{\Sigma_0} i\mathbf{k} P' - i\mathbf{k} \Phi', \tag{9.42}$$

where we note that here Φ' is evaluated on the plane $z = 0$.

We take an adiabatic equation of state of the form

$$P = K \Sigma^\Gamma, \tag{9.43}$$

where K is a constant and Γ is the adiabatic exponent appropriate for two dimensions. We also assume adiabatic perturbations so that

$$P' = C_s^2 \Sigma', \tag{9.44}$$

where the two-dimensional sound speed, C_s^2, is given by $C_s^2 = \Gamma P_0 / \Sigma_0$.

Using this, we now combine eqs. (9.40) and (9.42) to eliminate $\mathbf{u}$, and we obtain

$$-\omega^2 \Sigma' = k^2 C_s^2 \Sigma' + k^2 \Sigma_0 \Phi'. \tag{9.45}$$

We now need to use Poisson's equation to relate Φ' and Σ'. Because the density is confined to a thin slab, the linearized version of Poisson's equation is given by

$$\nabla^2 \Phi' = 4\pi G \Sigma' \delta(z), \tag{9.46}$$

or equivalently

$$\frac{d^2\Phi'}{dz^2} = k^2\Phi' + 4\pi G\Sigma'\delta(z).\tag{9.47}$$

Except on the plane $z = 0$, this has the following solution:

$$\Phi' = A\exp(-|kz|),\tag{9.48}$$

where A is a constant and we have imposed the condition that the perturbed potential vanishes at large $|z|$. To determine the value of A, we integrate eq. (9.46) with respect to z from $z = \epsilon$ to $z = -\epsilon$ and then take the limit $\epsilon \to 0$. This gives the jump in $d\Phi'/dz$ across the $z = 0$ plane as

$$\left[\frac{d\Phi'}{dz}\right]_{0^-}^{0^+} = 4\pi G\Sigma'.\tag{9.49}$$

Hence we conclude that

$$\Phi'(z = 0) = -\frac{2\pi G\Sigma'}{|k|}.\tag{9.50}$$

We now substitute this into eq. (9.45) to obtain the dispersion relation:

$$\omega^2 = k^2 C_s^2 - 2\pi G\Sigma_0|k|.\tag{9.51}$$

This has very similar properties to what we have found above. When gravity is negligible ($G = 0$) we just get acoustic waves with speed C_s. As gravity becomes more important, there is a critical value of the wavenumber k_J at which ω^2 changes sign and instability sets in. The wavenumber

$$k_J = \frac{2\pi G\Sigma_0}{C_s^2}\tag{9.52}$$

corresponds exactly to what we found in Section 9.2 for the isothermal gas layer.

However, once again there are problems with this analysis because the self-gravity of the equilibrium configuration in the (x, y)-direction is not properly accounted for. One way of providing a proper balance in the (x, y)-plane is to allow the configuration to rotate. This has obvious astrophysical applications. Rotation allows us to set up a balance between gravity and centrifugal force and to provide a fully self-consistent equilibrium. However, analysis of such a configuration will have to wait until we have looked at the analysis of rotating shear flows in Chapter 12.

9.4 Further reading

A simple description of the gravitational instability can be found in Jeans (1929, Chap. XIII). The stability of the isothermal self-gravitating plane layer is given by Ledoux (1951).

9.5 Problems

9.5.1 Show that a self-gravitating, isothermal slab of gas (sound speed c_s) centred on the plane $z = 0$ and immersed in a non-gravitating external medium with pressure p_e has a density profile $\rho_0(z)$, with midplane density ρ_{00}, where

$$\rho_0(\mu) = \rho_{00}(1 - \mu^2), \tag{9.53}$$

$$\mu = \tanh(z/h) \tag{9.54}$$

and

$$h = c_s/\sqrt{2\pi G \rho_{00}}, \tag{9.55}$$

valid for $-A < \mu < A$, where $A^2 = 1 - p_e/\rho_{00}c_s^2$.

If Σ is the surface density of the self-gravitating slab, show that

$$\rho_{00} = (2p_e + \pi G \Sigma^2)/2c_s^2. \tag{9.56}$$

9.5.2 Consider a *strongly* compressed isothermal slab, such that $p_e \approx \rho_{00}c_s^2 \gg G\Sigma^2/2$. In this approximation, the pressure and density are almost constant throughout the slab and the slab extends in the range $-a \le z \le a$, where $a = \Sigma/2\rho_{00} \ll h$. Consider a small perturbation of the slab with magnitude ξ, where $\xi/a \ll a/h \ll 1$. This implies that self-gravity may be ignored. Assume that the velocity of the perturbation is of the form $\mathbf{u} = (u, 0, w)$, that the perturbation is isothermal, so that $p' = c_s^2 \rho'$ and that the perturbed quantities have the following form:

$$p'(x, z, t) = p'(z) \exp\{i(\omega t - kx)\}. \tag{9.57}$$

Assume a fixed pressure boundary condition $p'(z = \pm a) = 0$. Show that the dispersion relation takes the form

$$\omega^2 = k^2 c_s^2 + \frac{N^2 \pi^2 c_s^2}{4a^2} \tag{9.58}$$

for all integers $N = 1, 2, 3, \ldots$

Sketch the group velocity against wavenumber for the fastest propagating mode. (See Doroshkevich (1980) and Lubow & Pringle (1993).)

9.5.3 A large spherical cloud of gas of radius R_0 is centred at the origin and has uniform density ρ_{00} and zero pressure . At time $t = 0$ it begins to collapse from rest under its own gravity. At time t the velocity field within the cloud may be written as follows:

$$\mathbf{u}_0(\mathbf{r}, t) = \mathbf{r} \left[\frac{\dot{R}(t)}{R(t)} \right], \tag{9.59}$$

where $R(t)$ is the cloud radius and $\mathbf{r}$ is the radius vector from the origin. Show that the density remains uniform and that at time t it is given by

$$\rho_0(t) = \rho_{00}[R_0/R(t)]^3. \tag{9.60}$$

Show that the gravitational force within the cloud is given by

$$\mathbf{g} = -\mathbf{r} \left[\frac{4}{3} \pi G \rho_0(t) \right] \tag{9.61}$$

and hence that

$$R^2\ddot{R} = -\frac{4}{3}\pi G\rho_{00}R_0^3. \tag{9.62}$$

Show that the collapse is described implicitly by

$$R = \frac{1}{2}R_0(1 + \cos\phi) \tag{9.63}$$

and

$$Ct = \frac{1}{2}R_0(\phi + \sin\phi), \tag{9.64}$$

where

$$C^2 = \frac{8}{3}\pi G\rho_{00}R_0^2. \tag{9.65}$$

As the cloud collapses (still with zero pressure) it is subject to small perturbations. Ignoring the effects of the cloud boundaries (or assuming the cloud is infinitely large), we assume that the perturbations take the form

$$\rho(\mathbf{r}, t) = \rho_0(t) + \rho'(\mathbf{r}, t), \tag{9.66}$$

with

$$\rho'(\mathbf{r}, t) = \rho_1(t)\exp(\mathrm{i}\mathbf{k} \cdot \mathbf{r}) \tag{9.67}$$

and where

$$\mathbf{k}(t) = \frac{\mathbf{q}}{R(t)} \tag{9.68}$$

and $\mathbf{q}$ is independent of t. Explain briefly the physical motivation for such an assumption.

Using this assumption, show that the equation of mass conservation implies

$$\dot{\rho}_1 + 3\rho_1\frac{\dot{R}}{R} + \mathrm{i}\frac{\rho_0}{R}\mathbf{q} \cdot \mathbf{u_1} = 0, \tag{9.69}$$

where $\mathbf{u}_1(t)$ is the analogous velocity perturbation.

Similarly, show that

$$\dot{\mathbf{u}}_1 + \frac{\dot{R}}{R}\mathbf{u_1} = -\mathrm{i}\frac{\mathbf{q}}{R}\Phi_1, \tag{9.70}$$

where Φ_1 is the analogous perturbation to the gravitational potential, and that

$$-\frac{q^2}{R^2}\Phi_1 = 4\pi G\rho_1. \tag{9.71}$$

Hence show that for compressive modes the fractional density perturbation, defined by $\delta(t) = \rho_1(t)/\rho_0(t)$, satisfies the following equation:

$$\ddot{\delta} + \frac{2\dot{R}}{R}\dot{\delta} - 4\pi G\rho_0(t)\delta = 0. \tag{9.72}$$

Show that one solution of this equation is given implicitly by

$$\delta(\phi) = \frac{\sin \phi}{(1 + \cos \phi)^2}. \tag{9.73}$$

By considering the behaviour of the solutions as $R \rightarrow 0$, i.e. as $\phi \rightarrow \pi$, or otherwise, show that this is the only growing solution as the collapse proceeds. (See Coles & Lucchin (1995) and Weinberg (1972).)

10

Linear shear flows

Shear flows are found in many areas of astronomy and occur whenever one fluid flows past another. An obvious place this occurs is when a jet flows into an ambient medium. A shear flow occurs at the sides of the jet and this is usually unstable, leading to energy release and radiative emission. Shear is also a characteristic feature of rotating discs. In this chapter we study the simplest case, where the shear flow is linear. Thus we consider flows whose unperturbed form in Cartesian coordinates is given by

$$\mathbf{U}(z) = (U(z), 0, 0). \tag{10.1}$$

We can see the strong tendency towards instability in such flows by considering a fluid of uniform density ρ lying between the two planes $z = \pm a$, whose velocity profile takes the form $U(z) = U_0(z/a)$, where U_0 is a constant. The total linear momentum of the fluid is zero. Thus, in principle, if there were some kind of instability which resulted in the fluid being completely mixed, the fluid would be brought to rest. This process would release an amount of energy given by

$$\Delta E = \int_{-a}^{a} \frac{1}{2}\rho U^2 \, \mathrm{d}z = \frac{1}{2}\rho U_0^2 \tag{10.2}$$

per unit length in the x-direction. This is a typical situation in which we can expect a fluid flow to be unstable. It occurs when some perturbation is able to tap an energy source of some kind (here the free shear energy) while obeying the conservation laws required by the fluid equations (here the conservation of linear momentum).

In this chapter we consider only incompressible fluids. This is a reasonable approximation if the relative shear motions are subsonic. This is in fact often not the case in astrophysical situations. However, if we included incompressibility we would have to consider the added complication that the shear layer can lose energy by emitting acoustic waves. For the sake of simplicity we shall ignore this aspect of the problem.

10.1 Perturbation of a linear shear flow

We consider initially an incompressible fluid of uniform density ρ. Then the mass conservation equation is simply

$$\text{div } \mathbf{u} = 0 \tag{10.3}$$

and the momentum equation is given by

$$\frac{\partial \mathbf{u}}{\partial t} + (\mathbf{u} \cdot \nabla)\mathbf{u} = -\nabla P, \tag{10.4}$$

where for convenience we define $P = p/\rho$.

The unperturbed flow has the form

$$\mathbf{U} = (U(z), 0, 0), \tag{10.5}$$

and $P = P_0 = \text{const}$. We take the flow to lie between fixed boundaries at $z = z_1$ and $z = z_2$, with $z_1 < z_2$.

In the perturbed flow the velocity has the form $\mathbf{U} + \mathbf{u}$, where

$$\mathbf{u} = (u, v, w), \tag{10.6}$$

and the pressure is now $P = P_0 + P'$. The perturbed mass conservation is then given by

$$\frac{\partial u}{\partial x} + \frac{\partial v}{\partial y} + \frac{\partial w}{\partial z} = 0, \tag{10.7}$$

and the components of the perturbed momentum equation are as follows:

$$\frac{\partial u}{\partial t} + U(z)\frac{\partial u}{\partial x} + w\frac{dU}{dz} = -\frac{\partial P'}{\partial x}, \tag{10.8}$$

$$\frac{\partial v}{\partial t} + U(z)\frac{\partial v}{\partial x} = -\frac{\partial P'}{\partial y} \tag{10.9}$$

and

$$\frac{\partial w}{\partial t} + U(z)\frac{\partial w}{\partial x} = -\frac{\partial P'}{\partial z}. \tag{10.10}$$

Using the symmetries of the unperturbed configuration we can now Fourier transform with respect to x, y and t, but we need to retain the full dependence in z. Thus we take all quantities to vary as $\exp\{i(\omega t - k_x x - k_y y)\}$. We shall also denote the derivative of U as $U' = dU/dz$.

Then eqs. (10.7)–(10.10) become

$$-(\mathrm{i}k_x u + \mathrm{i}k_y v) + \frac{dw}{dz} = 0, \tag{10.11}$$

$$\mathrm{i}(\omega - k_x U)u + U'w = \mathrm{i}k_x P', \tag{10.12}$$

$$\mathrm{i}(\omega - k_x U)v = \mathrm{i}k_y P' \tag{10.13}$$

and

$$i(\omega - k_x U)w = -\frac{dP'}{dz}. \tag{10.14}$$

To these we must add the boundary conditions that $w = 0$ at $z = z_1, z_2$.

10.2 Squire's theorem

We can simplify the above analysis without loss of generality. We define $k^2 = k_x^2 + k_y^2$ and $\tilde{u}$ by

$$k\tilde{u} = k_x u + k_y v. \tag{10.15}$$

Then eq. (10.11) can be written as follows:

$$-ik\tilde{u} + \frac{dw}{dz} = 0. \tag{10.16}$$

Similarly by multiplying eq. (10.12) by k_x/k, eq. (10.13) by k_y/k and adding, these equations combine to yield

$$i(\omega - k_x U)\tilde{u} + \frac{k_x}{k}U'w = ikP'. \tag{10.17}$$

Defining $\tilde{U} = k_x U/k$, this can be further simplified to

$$i(\omega - k\tilde{U})\tilde{u} + \tilde{U}'w = ikP'. \tag{10.18}$$

Also, eq. (10.14) becomes

$$i(\omega - k\tilde{U})w = -\frac{dP'}{dz}. \tag{10.19}$$

By comparing eq. (10.16) with eq. (10.11), eq. (10.18) with eqs. (10.12) and (10.13) and eq. (10.19) with eq. (10.14), we see that the set of equations we have just derived is equivalent to the original set with the transformations $k_x \to k$, $k_y \to 0$, $u \to \tilde{u}$, $v \to 0$ and $U \to \tilde{U}$. This implies that as far as the instability of the shear layer is concerned, we can take $k_y = 0$ and $v = 0$ without loss of generality, and need therefore only consider two-dimensional disturbances in the (x, z)-plane. This is Squire's theorem.

10.3 Rayleigh's inflexion point theorem

Our equations describing the perturbed flow are now as follows:

$$-iku + \frac{dw}{dz} = 0, \tag{10.20}$$

$$i(\omega - kU)u + U'w = ikP' \tag{10.21}$$

and

$$i(\omega - kU)w = -\frac{dP'}{dz}. \tag{10.22}$$

We now write these as a single differential equation for w. We first eliminate u between eqs. (10.20) and (10.21) to obtain

$$\frac{\omega - kU}{k}\frac{dw}{dz} + U'w = ikP'. \tag{10.23}$$

We then differentiate this with respect to z, remembering that U is a function of z, and use eq. (10.21) to eliminate dP'/dz. We then obtain what is known as Rayleigh's equation

$$\left(\frac{\omega}{k} - U\right)\left(\frac{d^2w}{dz^2} - k^2w\right) + \frac{d^2U}{dz^2}w = 0. \tag{10.24}$$

This equation is subject to the boundary conditions $w = 0$ at $z = z_1, z_2$.

If we assume that k is real, so that we are concerned with travelling modes rather than evanescent ones, and if we write

$$\omega = \omega_R + i\omega_I, \tag{10.25}$$

then the solutions are stable if $\omega_I > 0$ and unstable if $\omega_I < 0$. The equation is unchanged by the transformations $\omega \to -\omega$ and $k \to -k$. This means that we can, without loss of generality, take k to be real with $k > 0$. In addition, if ω is the eigenvalue with eigenfunction w, then the complex conjugate ω^* is also an eigenvalue with eigenfunction w^*. This implies that to prove instability all we need to do is show that $\omega_I \neq 0$.

Rayleigh's inflexion point theorem then states that: a *necessary* condition for instability is that the velocity profile has an inflexion point, i.e. $d^2U/dz^2 = 0$ for some value of $z = z_*$ in the range $z_1 < z_* < z_2$. To prove this we rewrite eq. (10.24) as follows:

$$\frac{d^2w}{dz^2} - k^2w - \frac{d^2U/dz^2}{U - (\omega/k)}w = 0. \tag{10.26}$$

We assume that $\omega_I \neq 0$ so that the equation is non-singular. We then multiply it by w^* and integrate over the fluid. Thus,

$$\int_{z_1}^{z_2}\left(\left|\frac{dw}{dz}\right|^2 + k^2|w|^2\right)dz + \int_{z_1}^{z_2}\frac{d^2U/dz^2}{U - (\omega/k)}|w|^2\,dz = 0. \tag{10.27}$$

Taking the imaginary part of this equation, we find that

$$\omega_I\int_{z_1}^{z_2}\frac{d^2U/dz^2}{|U - (\omega/k)|^2}|w|^2\,dz = 0. \tag{10.28}$$

Then, in order for ω_I to be non-zero, we require that the integral vanish. This can only happen if d^2U/dz^2 is positive for some values of z and negative for others.

This means that there must be some point $z = z_*$ at which $d^2U/dz^2 = 0$, i.e. there must be an inflexion point in the velocity profile.

10.3.1 Mathematical technicality

Equation (10.26) is formally singular for values of $\omega = kU(z)$ with $z_1 < z < z_2$. This implies that when we come to carry out formally the contour integration to invert the Fourier transform and find the real solution as a function of time t, we may not only have to evaluate the function at poles, which typically give rise to exponential behaviour, but also have to evaluate the integral along a cut on the real ω-axis. When using Fourier transforms there is then some ambiguity as to which side of the cut the contour should be drawn. For this reason it often makes more physical sense to think in terms of solving the problem as an initial-value problem and so to use Laplace transforms. In this case it is clear where the inverse contour should be drawn. One then ends up with contour integrals around a cut (or cuts) on the imaginary axis. Such contour integrals typically give rise to algebraic, rather than exponential, time behaviour. We note here that in the above analysis we have implicitly assumed that any instability has exponential behaviour, i.e. that the contributions from the poles in the complex ω-plane dominate. We have, however, not proved this, and so we have not discussed the stability or otherwise of the 'singular modes' which correspond to the cut along the real ω-axis.

10.4 Fjørtoft's theorem

There is a slightly more stringent necessary condition for instability known as Fjørtoft's theorem. This states that a necessary condition for instability is that if z_* is a point at which $d^2U/dz^2 = 0$, then there must be some value z_0 in the range $z_1 < z_0 < z_2$ such that

$$\left.\frac{d^2U}{dz^2}\right|_{z_0} [U(z_0) - U(z_*)] < 0. \tag{10.29}$$

We prove this as follows. Take the real part of eq. (10.27). This yields

$$\int_{z_1}^{z_2} \left(\frac{d^2U}{dz^2}\right) \frac{U - (\omega_R/k)}{|U - \omega/k|^2} |w|^2 \, dz = -\int_{z_1}^{z_2} \left(\left|\frac{dw}{dz}\right|^2 + k^2|w|^2\right) dz < 0. \tag{10.30}$$

Then, if we have instability ($\omega_I \neq 0$), we can add to the l.h.s. of this equation a quantity which from eq. (10.28) we know to be zero, namely

$$[(\omega_R/k) - U(z_*)] \int_{z_1}^{z_2} \frac{d^2U/dz^2}{|U - \omega/k|^2} |w|^2 \, dz = 0. \tag{10.31}$$

When we do this, we obtain the following result:

$$\int_{z_1}^{z_2} \left(\frac{d^2 U}{dz^2} \right) \frac{U(z) - U(z_*)}{|U - (\omega/k)|^2} |w|^2 \, dz < 0, \tag{10.32}$$

which proves the theorem.

10.5 Physical interpretation

On the face of it, Rayleigh's theorem and Fjørtoft's theorem look like random results which have emerged by chance from the mathematics. But of course there must be some physical reasons behind the results. The simplest way to think about these results is in terms of vorticity. For the unperturbed flow $(U(z), 0, 0)$, the vorticity is of the form $\boldsymbol{\omega} = \nabla \wedge \mathbf{u} = (0, \omega(z), 0)$, where

$$\omega(z) = \frac{dU}{dz}. \tag{10.33}$$

In addition we are considering perturbations of the form

$$\mathbf{u} = (u, 0, w). \tag{10.34}$$

In an incompressible fluid the equations of motion imply an equation for the evolution of vorticity which takes the following form:

$$\frac{D\boldsymbol{\omega}}{Dt} = (\boldsymbol{\omega} \cdot \nabla)\mathbf{u}. \tag{10.35}$$

This implies that the vortex lines all lie parallel to the y-axis and are moved around in the (x, z)-plane. Thus they are all conserved in strength.

We have argued that in order to release free shear energy , and so produce an instability, we need to interchange fluid elements in an appropriate manner. Rayleigh's criterion tells us that in order to produce an instability there must be a point at which $d^2 U / dz^2 = 0$, or equivalently there must be a point at which $d\omega/dz = 0$. At this point there are neighbouring fluid elements which have the same vorticity but different velocities. Mixing such fluid elements, and so releasing shear energy, is permitted by the fluid equations. Thus if $d\omega/dz$ is zero at some point, we have the possibility of an instability.

In Figs. 10.1(a) and (b) we show two velocity profiles which do not have an inflexion point and so are stable under these considerations. In Figs. 10.1(c) and (d) we show two velocity profiles which do have inflexion points, and so, by Rayleigh's criterion, might show instability. The profile in Fig. 10.1(c), however, does not obey the necessary condition given by Fjørtoft's theorem, and so is stable. In contrast, the profile in Fig. 10.1(d) does obey the additional criterion and so might be unstable. What is the basic physical difference between these two? Here again it is illuminating to consider the corresponding vorticity profiles $\omega(z)$ (Fig. 10.2).

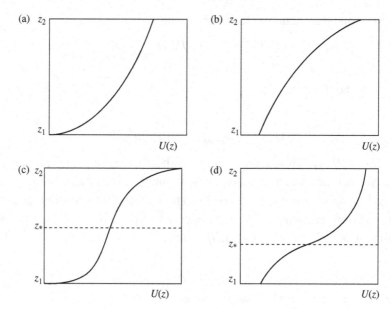

Fig. 10.1. Velocity profiles illustrating Fjørtoft's theorem. (a), (b) These profiles have no inflexions, and so are stable by Rayleigh's criterion. (c) A profile which has an inflexion, but does not fulfil the conditions of Fjørtoft's theorem, and so is again stable. (d) A profile which does fulfil the conditions of the theorem, and so can be unstable.

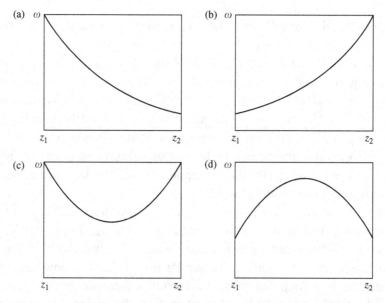

Fig. 10.2. The vorticity profiles of the flows shown in Fig. 10.1; profile (d) has a vorticity maximum. Thus mixing reduces the vorticity, releasing energy and allowing the possibility of instability.

The energy which drives a shear instability derives from reducing the local shear, that is from reducing the local vorticity. At the same time, we know that from the vorticity equation the total vorticity of the fluid must be conserved. Moreover, the effect of mixing neighbouring fluid elements is, on a coarse scale, to smooth out any vorticity profile. The profile shown in Fig. 10.1(c) has a vorticity profile with a minimum at $z = z_*$. The effect of mixing here is therefore to increase the vorticity at z_*. This does not release shear energy and so does not drive an instability. In contrast, the profile shown in Fig. 10.1(d) has a vorticity profile with a maximum at $z = z_*$. The effect of mixing here is to reduce the local vorticity and so release shear energy. Thus for this profile we have the possibility of a shear-driven instability.

10.6 Co-moving phase

We have so far looked at the properties the velocity profile of the fluid must have in order that there might be an instability. Here we look at the properties of the unstable mode. We start with Rayleigh's equation (eq. (10.24)) in the form

$$\left(U - \frac{\omega}{k}\right) \frac{d^2 w}{dz^2} - \left[k^2 \left(U - \frac{\omega}{k}\right) + \frac{d^2 U}{dz^2}\right] w = 0. \tag{10.36}$$

We now introduce a new variable ψ such that

$$\psi = \frac{w}{U - (\omega/k)}. \tag{10.37}$$

Then Rayleigh's equation becomes

$$\frac{d}{dz}\left[\left(U - \frac{\omega}{k}\right)^2 \frac{d\psi}{dz}\right] - k^2 \left(U - \frac{\omega}{k}\right)^2 \psi = 0. \tag{10.38}$$

We note in passing that this equation corresponds to a Sturm–Liouville problem in that it comes from extremizing an integral in the form

$$\delta \int_{z_1}^{z_2} \left(U - \frac{\omega}{k}\right)^2 \left[\left(\frac{d\psi}{dz}\right)^2 + k^2 \psi^2\right] dz = 0. \tag{10.39}$$

Returning to the main argument, we multiply eq. (10.38) by ψ^* and integrate over the fluid to give

$$\int_{z_1}^{z_2} \left(U - \frac{\omega}{k}\right)^2 \left[\left|\frac{d\psi}{dz}\right|^2 + k^2 |\psi|^2\right] dz = 0. \tag{10.40}$$

Taking the imaginary part of this equation, we then obtain

$$\omega_I \int_{z_1}^{z_2} \left(U - \frac{\omega_R}{k}\right) \left[\left|\frac{d\psi}{dz}\right|^2 + k^2 |\psi|^2\right] dz = 0. \tag{10.41}$$

We now see that for an unstable mode, for which we require $\omega_I \neq 0$, there must be a point z_p such that $z_1 < z_p < z_2$ and $\omega_R/k = U(z_p)$. This means that an unstable mode must have a phase velocity equal to the fluid velocity at some point. The mode must co-move with some of the fluid. This also implies that some parts of the fluid move faster than the mode and some parts move slower. Thus, from a physical point of view, the mode 'knows' about the shear and is in a position to allow communication between faster and slower moving fluid elements. It is this communication which allows the mode to tap the shear energy, and so to grow.

10.7 Stratified shear flow

We now consider the situation in which a shear flow moves horizontally in a vertical gravitational field. Thus, as before, the unperturbed flow takes the following form:

$$\mathbf{U} = (U(z), 0, 0), \tag{10.42}$$

but now the unperturbed fluid has a density gradient $\rho(z)$ and is subject to a uniform gravitational acceleration $\mathbf{g} = (0, 0, -g)$. For convenience we still assume that the fluid is incompressible. We take the velocity perturbation as (u, v, w).

From a physical point of view, the instability is now controlled by two physical effects. First, there is the free energy of the shear flow, which can be tapped to produce growing modes. Second, there is the energy available from the gravitational field. In particular, if the fluid is stably stratified (here this implies $d\rho/dz < 0$, so that the heavier fluid is at the bottom) then the vertical mixing required to tap the shear energy needs to give up some of its energy to gravitation. Thus we expect that instability in this case arises as a balance between shear energy and gravitational energy.

As before, we Fourier transform the perturbation equations in the form $\propto \exp\{i(\omega t + k_x x + k_y y)\}$. The mass conservation equation div $\mathbf{u} = 0$ yields

$$ik_x u + ik_y v + \frac{dw}{dz} = 0. \tag{10.43}$$

Alternatively we can use $D\rho/Dt = 0$ to yield

$$i(\omega + k_x U)\rho' + w\frac{d\rho}{dz} = 0. \tag{10.44}$$

These two equations are equivalent. Note that although the Lagrangian density perturbation $\delta\rho$ is zero, the Eulerian density perturbation ρ' is not. This is because there is a non-zero density gradient in the unperturbed fluid. Previously both $d\rho/dz$ and ρ' were zero, and so the second equation was trivially satisfied.

The three components of the linearized momentum equations are given by

$$i(\omega + k_x U)\rho u + \rho \frac{dU}{dz} w = -ik_x p',$$ (10.45)

$$i(\omega + k_x U)\rho v = -ik_y p'$$ (10.46)

and

$$i(\omega + k_x U)\rho w = -\frac{dp'}{dz} - g\rho'.$$ (10.47)

We now obtain an expression for u from eq. (10.45), namely

$$u = -\frac{k_x p'}{(\omega + k_x U)\rho} - \frac{wU'}{i(\omega + k_x U)},$$ (10.48)

and an expression for v from eq. (10.46), namely

$$v = -\frac{k_y p'}{(\omega + k_x U)\rho},$$ (10.49)

and substitute them into eq. (10.43) to obtain an expression for p' in terms of w and its derivative:

$$p' = -i\frac{(\omega + k_x U)\rho}{k_x^2 + k_y^2}\frac{dw}{dz} + i\frac{k_x w\rho U'}{k_x^2 + k_y^2}w.$$ (10.50)

We also eliminate ρ' between eqs. (10.47) and (10.44) to obtain an equation for dp'/dz in terms of w as follows:

$$\frac{dp'}{dz} = -i(\omega + k_x U)\rho w - i\frac{g(d\rho/dz)}{\omega + k_x U}w.$$ (10.51)

Combining eqs. (10.50) and (10.51) now gives a second-order differential equation for w as follows:

$$\frac{d}{dz}\left\{(\omega + k_x U)\rho\frac{dw}{dz} - k_x \rho U'w\right\} = k_\perp^2(\omega + k_x U)\rho w + \frac{k_\perp^2 g(d\rho/dz)}{\omega + k_x U}w,$$ (10.52)

where we have written $k_\perp^2 = k_x^2 + k_y^2$.

We now use physical intuition to make an approximation which, while not correct in general, is justified for the purpose we have in mind. What we are looking for here is not the general dynamical behaviour of the perturbed shear flow, but rather a stability criterion which gives us information about where the border between stability and instability lies. If we expand the l.h.s. of eq. (10.52) we obtain a number of terms which contain the derivative $d\rho/dz$. Now this derivative is related (eq. (10.46)) to the Eulerian density perturbation ρ', and the change in local density has two physical effects. First, it changes the local inertia of the fluid; that is, it changes the timescale on which the fluid reacts to a given force. Second, it changes the local buoyancy of the fluid; that is, it changes the stability of the fluid. Since we are only interested in stability, and not in exact dynamical timescales, we can neglect the variation of ρ except where it is coupled to the gravity g. In particular, when expanding the derivative on the l.h.s. of eq. (10.52) we can treat ρ as a constant.

This procedure yields the following equation:

$$(\omega + k_x U) \left\{ \frac{d^2 w}{dz^2} - k_\perp^2 w \right\} - k_x \frac{d^2 U}{dz^2} w - \frac{k_\perp^2}{\omega + k_x U} \left(\frac{g}{\rho} \frac{d\rho}{dz} \right) w = 0. \quad (10.53)$$

As before, we now simply consider two-dimensional perturbations, so we take $k_\perp = k_x = k$, and for convenience we define the phase velocity of the mode as $c = \omega/k$. Then the equation becomes

$$(c + U) \left(\frac{d^2 w}{dz^2} - k^2 w \right) - \frac{d^2 U}{dz^2} w + \left(\frac{-g}{\rho} \frac{d\rho}{dz} \right) \frac{w}{c + U} = 0. \quad (10.54)$$

This is called the Taylor–Goldstein equation.

10.8 The Richardson criterion

We are now in a position to obtain a *necessary* condition for the instability of this stratified shear flow in a gravitational field.

We first define a quantity H as follows:

$$H = \frac{w}{(c + U)^{1/2}}. \quad (10.55)$$

In terms of this variable, eq. (10.54) becomes

$$\frac{d}{dz} \left[(U + c) \frac{dH}{dz} \right] - k^2 (c + U) H - \frac{1}{2} \frac{d^2 U}{dz^2} H - \left\{ \frac{1}{4} U'^2 - \left[\frac{g}{\rho} \left(-\frac{d\rho}{dz} \right) \right] \right\} \frac{H}{c + U} = 0. \quad (10.56)$$

Note that for the fluid to be stably stratified in the absence of shear, we require that $-d\rho/dz > 0$.

We now multiply this equation by the complex conjugate H^* and integrate over the fluid (from z_1 to z_2). We integrate by parts where necessary, using the boundary conditions that $w = 0$ at $z = z_1, z_2$. We then obtain the following equation:

$$\int_{z_1}^{z_2} \left[\left\{ \left| \frac{dH}{dz} \right|^2 + k^2 |H|^2 \right\} + \frac{1}{2} \frac{d^2 U}{dz^2} |H|^2 + \left\{ \frac{1}{4} U'^2 - \left[\frac{g}{\rho} \left(-\frac{d\rho}{dz} \right) \right] \right\} \frac{|H|^2}{c + U} \right] dz = 0. \quad (10.57)$$

Now we write c in terms of real and imaginary parts,

$$c = c_R + ic_I, \tag{10.58}$$

and note, as before, that for instability we just require that $c_I \neq 0$. Therefore we take the imaginary part of eq. (10.57), which is as follows:

$$c_I \int_{z_1}^{z_2} \left[\left| \frac{dH}{dz} \right|^2 + k^2 |H|^2 + \frac{\{g/\rho \, (-d\rho/dz) - U'^2/4\}}{|c + U|^2} \right] dz = 0. \tag{10.59}$$

We see by inspection that if throughout the fluid, i.e. in $z_1 < z < z_2$, we have

$$\frac{1}{4} U'^2 < \left[\frac{g}{\rho} \left(-\frac{d\rho}{dz} \right) \right], \tag{10.60}$$

then the integrand is positive definite; therefore to satisfy the equation we require $c_I = 0$. Thus if the inequality eq. (10.60) is satisfied throughout the fluid, then the shear flow must be stable. This is known as the Richardson criterion. Physically speaking, the l.h.s. is a measure of the energy available in the shear and the r.h.s. is a measure of the energy required to overturn the density gradient in the presence of gravity. Thus the criterion states roughly that if there is not enough shear energy to overturn the density gradient then the fluid is stable. Note that the criterion only goes one way. It does not demonstrate instability if the inequality is not satisfied.

10.9 Further reading

A detailed discussion of the theory of the stability of incompressible shear flows is to be found in Drazin & Reid (1981, Chap. 4). Basic results about the Richardson criterion are given by Howard (1961) and Miles (1961). Consideration of the effects of buoyancy, surface tension and magnetic fields can be found in Chandrasekhar (1961, Chap. XI). The complications introduced by consideration of compressibility and the emission of acoustic waves from the shear layer are discussed by Gerwin (1968).

10.10 Problems

10.10.1 A smooth pipe with square cross section of side a lies flat on a horizontal surface in the uniform gravitational field g. The pipe contains incompressible fluid.

(i) Consider the case where the fluid in the upper half of the pipe has density ρ_1 and that in the lower half-pipe has density ρ_0, where $\rho_1 > \rho_0$. What is the physical origin of the instability which ensues? Calculate the energy E_1 available to drive the instability per unit length of the pipe.

(ii) Now consider the case where $\rho_1 < \rho_0$, but with the fluid in the upper half-pipe moving with velocity V and the fluid in the lower half-pipe moving with velocity $-V$. If we neglect gravity, setting $g = 0$, what is the physical origin of the instability which ensues? Show that the energy available to the instability, per unit length of pipe, is given by

$$E_0 = \frac{\rho_0 \rho_1}{\rho_0^2 + \rho_1^2} a^2 V^2. \tag{10.61}$$

(iii) If now $\rho_1 < \rho_0$, but $g \neq 0$, explain by comparing E_0 and E_1 the physical significance of the Richardson number, Ri, where

$$\text{Ri} = \frac{\rho_0 \rho_1}{\rho_0^2 - \rho_1^2} \frac{V^2}{ag}. \tag{10.62}$$

10.10.2 Consider a horizontal flow of incompressible fluid in a vertical, constant gravitational field g, so that

$$\mathbf{u} = \begin{cases} (U_2, 0, 0), & z > 0, \\ (U_1, 0, 0), & z < 0, \end{cases} \tag{10.63}$$

$$\rho = \begin{cases} \rho_2, & z > 0, \\ \rho_1, & z < 0 \end{cases} \tag{10.64}$$

and

$$p(z) = \begin{cases} p_0 - g\rho_2 z, & z > 0, \\ p_0 - g\rho_1 z, & z < 0. \end{cases} \tag{10.65}$$

Consider small perturbations, such that the interface of the fluid is at $z = \zeta(x, y, t)$, which leave the flow unperturbed as $|z| \to \infty$. Assume that the perturbed flow is irrotational (zero vorticity) so that, for example, the flow in $z > 0$ can be written in terms of a velocity potential $\mathbf{u} = \nabla\phi_2$, where $\phi_2 = U_2 x + \phi_2'$ and $\nabla^2 \phi_2' = 0$.

Show that the velocity potential obeys the following equation:

$$\frac{\partial\phi}{\partial t} + \frac{1}{2}(\nabla\phi)^2 + \frac{p}{\rho} + gz = \text{const.} \tag{10.66}$$

Take the perturbed quantities to be of the form

$$\phi_2' = \hat{\phi}_2(z) \exp\{i(kx + \omega t)\} \tag{10.67}$$

and find expressions for $\hat{\phi}_1$ and $\hat{\phi}_2$.

Show that the dispersion relation of the perturbations is given by

$$\omega^2(\rho_1 + \rho_2) + 2\omega k(\rho_2 U_2 + \rho_1 U_1) + k^2(\rho_2 U_2^2 + \rho_1 U_1^2) + (\rho_2 - \rho_1)kg = 0. \tag{10.68}$$

Discuss what happens when $U_1 = U_2 = 0$ in the two cases (i) $\rho_1 < \rho_2$, (ii) $\rho_1 > \rho_2$.

Show that if $U_1 \neq U_2$ the flow is always unstable for perturbations of short enough wavelength. (See Drazin & Reid (1981, Chap. 1).)

10.10.3 An incompressible fluid of uniform density ρ has a uniform magnetic field (in Cartesian coordinates) $\mathbf{B}_0 = (B, 0, 0)$ and a shearing velocity field $\mathbf{u}_0 = (U(z), 0, 0)$.

The flow is subject to small perturbations of the form

$$\mathbf{u} = \mathbf{u}_0 + [u(z), v(z), w(z)] \exp\{i(\omega t + k_x x + k_y y)\}, \tag{10.69}$$

$$\mathbf{B} = \mathbf{B}_0 + [b_x(z), b_y(z), b_z(z)] \exp\{i(\omega t + k_x x + k_y y)\} \tag{10.70}$$

and

$$p = p_0 + p_1(z) \exp\{i(\omega t + k_x x + k_y y)\} \tag{10.71}$$

From the linearized equation of motion, show that

$$i\rho(\omega + k_x U)w - B\left(ik_x b_z - \frac{db_x}{dz}\right) = -\frac{dp_1}{dz} \tag{10.72}$$

and derive the corresponding equations for u and v.

From the linearized induction equation, show that

$$b_x = \frac{k_x B}{\omega + k_x U} \left\{ u - \frac{iU'w}{\omega + k_x U} \right\}, \tag{10.73}$$

where $U' = dU/dz$, and obtain analogous expressions for b_y and b_z in terms of the perturbed velocity components.

Substitute these expressions for the perturbed components of the magnetic field into the linearized equations of motion.

From the x- and y-components of the linearized equations of motion, show that the z-component of the vorticity, $\zeta = ik_x v - ik_y u$, is given by

$$\zeta = \frac{k_y U' w}{\omega + k_x U}. \tag{10.74}$$

Deduce that the y-component of the linearized equation of motion simplifies to

$$i\rho(\omega + k_x U)v = -ik_y p_1. \tag{10.75}$$

Use div $\mathbf{u} = 0$ to show that

$$ik^2 p_1 = \rho(\omega + k_x U)\frac{dw}{dz} - \rho k_x U' w, \tag{10.76}$$

where $k^2 = k_x^2 + k_y^2$.

Combine div $\mathbf{u} = 0$ with the expression for ζ to show that

$$ik^2 u = -\left[k_x \frac{dw}{dz} + \frac{k_y^2 U'}{\omega + k_x U} w \right]. \tag{10.77}$$

Hence, or otherwise, obtain an equation for w in the following form:

$$\frac{d}{dz}\left\{ \rho(\omega + k_x U)\frac{dw}{dz} - \rho k_x U' w \right\} = k^2 \rho(\omega + k_x U)w$$

$$+ k_x^2 B^2 \left\{ \frac{d}{dz}\left(\frac{dw/dz}{\omega + K_x U} \right) - \frac{k^2 w}{\omega + k_x U} \right\} - k_x^3 B^2 \frac{d}{dz}\left\{ \frac{U'w}{(\omega + k_x U)^2} \right\}.$$

Consider a shear layer at $z = 0$, so that

$$U(z) = \begin{cases} U_2, & z > 0, \\ U_1, & z < 0, \end{cases} \qquad (10.78)$$

where U_1 and U_2 are constants. Show that in this case the solutions of eq. (10.78) for which $w/(\omega + k_x U)$ is continuous and obey suitable boundary conditions as $z \to \pm\infty$ are given by

$$w = \begin{cases} A(\omega + k_x U_2)e^{-kz}, & z > 0, \\ A(\omega + k_x U_1)e^{kz}, & z < 0, \end{cases} \qquad (10.79)$$

for some constant A and for $k > 0$.

By integrating eq. (10.78) from $z = -\epsilon$ to $z = \epsilon$ and letting $\epsilon \to 0$, show that

$$\rho(\omega + k_x U_2)^2 + \rho(\omega + k_x U_1)^2 = 2k_x^2 B^2. \qquad (10.80)$$

Deduce that the shear flow is stable if $(U_1 - U_2)^2 < 4B^2/\rho$. Why does the presence of this magnetic field tend to stabilize the shear flow? (See Chandrasekhar (1961, Chap. XI).)

10.10.4 A uniform density, incompressible fluid contains a bounded shear layer with velocity field, in Cartesian coordinates, of the form $\mathbf{U}_0 = (U(z), 0, 0)$, where

$$U(z) = \begin{cases} U_0, & z \geq d, \\ (z/d)U_0, & |z| < d, \\ -u_0, & z \leq -d. \end{cases} \qquad (10.81)$$

It is subject to a perturbation of the form

$$\mathbf{u}(x, z, t) = (u(z), 0, w(z)) \exp\{ik(x - ct)\} \qquad (10.82)$$

and

$$p'(x, z, t) = p'(z) \exp\{ik(x - ct)\}, \qquad (10.83)$$

where $k > 0$ is real. Obtain an equation for $w(z)$ and show that an appropriate solution is of the form

$$w(z) = \begin{cases} A \exp[-k(z - d)], & z \geq d, \\ B \exp[-k(z - d)] + C \exp[k(z + d)], & |z| < d, \\ D \exp[k(z + d)], & z \leq -d, \end{cases} \qquad (10.84)$$

where A, B, C and D are constants.

Explain why the quantity

$$\left[(c - U)\frac{dw}{dz} + \frac{dU}{dz}w \right] \qquad (10.85)$$

is continuous at $z = \pm d$.

Hence, or otherwise, show that

$$\left(\frac{c}{U_0}\right)^2 = \frac{1}{4\alpha^2}\{(2\alpha - 1)^2 - e^{-4\alpha}\}, \tag{10.86}$$

where $\alpha = kd$.

Deduce that disturbances with wavelengths such that $0 < \alpha < \alpha_s$ are unstable for some number α_s, which is in the range $\frac{1}{2} < \alpha_s < 1$.

Give a physical interpretation of this result. (See Drazin & Reid (1981, Chap. 23).)

11

Rotating flows

Almost all astronomical objects rotate. The formation of astronomical objects often involves gravitational collapse over many orders of magnitude in size. Conservation of angular momentum then ensures that newly formed objects often rotate rapidly in the sense that rotational (centrifugal) forces play a dynamical role. In this and the next few chapters we consider the effects of rotation. In doing so we shall have in mind mainly the effects of rotation on stellar objects, but note that some of our findings are also relevant to other types of object.

11.1 Rotating fluid equilibria

We begin by asking what kinds of equilibrium are possible for a rotating self-gravitating fluid mass. We consider a fluid rotating at a steady rate about an axis of symmetry. In cylindrical polar coordinates (R, ϕ, z), the flow velocity is given by

$$\mathbf{u} = (0, u_\phi, 0), \tag{11.1}$$

where $u_\phi = R\Omega$ is independent of the azimuthal coordinate ϕ and Ω is the angular velocity. Since the flow is axisymmetric ($\partial/\partial\phi = 0$), there are only two non-zero components of the equation of hydrostatic equilibrium. Denoting the gravitational potential as Φ, the R-component is given by

$$-\frac{u_\phi^2}{R} = -\frac{1}{\rho}\frac{\partial p}{\partial R} - \frac{\partial \Phi}{\partial R} \tag{11.2}$$

and the z-component is given by

$$0 = -\frac{1}{\rho}\frac{\partial p}{\partial z} - \frac{\partial \Phi}{\partial z}. \tag{11.3}$$

This can be written in differential form as follows:

$$\frac{\mathrm{d}p}{\rho} = g_R\,\mathrm{d}R + g_z\,\mathrm{d}z, \tag{11.4}$$

where the vector **g** is given by

$$\mathbf{g} = \left(-\frac{\partial \Phi}{\partial R} + R\Omega^2, 0, -\frac{\partial \Phi}{\partial z} \right). \tag{11.5}$$

The r.h.s. of this equation is the gradient of a scalar if and only if

$$\frac{\partial g_z}{\partial R} = \frac{\partial g_R}{\partial z}, \tag{11.6}$$

which is true if and only if

$$\frac{\partial \Omega}{\partial z} = 0. \tag{11.7}$$

If this holds, so that Ω is a function of R only, we may define an *effective* gravitational potential Φ_e as follows:

$$\Phi_e = \Phi(R, z) - \int^R \Omega^2 R \, dR. \tag{11.8}$$

Then the effective gravity $\mathbf{g}_e$ can be defined as

$$\mathbf{g}_e = -\nabla \Phi_e, \tag{11.9}$$

and the equation of hydrostatic equilibrium becomes

$$\frac{\nabla p}{\rho} = \mathbf{g}_e. \tag{11.10}$$

We note in passing that if the rotation is uniform, i.e. $\Omega = \Omega_0 = \text{const.}$, then

$$\Phi_e = \Phi - \frac{1}{2} R^2 \Omega_0^2. \tag{11.11}$$

Returning to the general case, $\Omega = \Omega(R)$, we see that hydrostatic equilibrium, expressed as

$$\nabla p = -\rho \nabla \Phi_e, \tag{11.12}$$

means that ∇p and $\nabla \Phi_e$ are everywhere parallel. Thus p must be a function of Φ_e, i.e. $p = p(\Phi_e)$. Then, since $\nabla p / \rho = -\nabla \Phi_e$, we have that $\rho = dp/d\Phi_e$ is also just a function of Φ_e. Hence we must have $p = p(\rho)$, so the fluid is *barotropic* . We have shown therefore that

$$\Omega = \Omega(R) \Leftrightarrow p = p(\rho). \tag{11.13}$$

11.2 Making rotating stellar models

The work in Section 11.1 shows that our freedom in building a model of a rotating star is severely reduced when we try to make simplifying assumptions. We have seen that to make a star which rotates uniformly (or one which rotates on cylinders) hydrostatic equilibrium requires the fluid to be barotropic. In addition, both pressure p and density ρ are functions only of the effective gravitational potential Φ_e. If as usual the stellar gas obeys a perfect gas equation of state, we have $p \propto \rho T$ and the temperature also is just a function of Φ_e.

The energy equation is given by

$$\rho T \frac{Ds}{Dt} = \rho \epsilon_{nuc} - \text{div } \mathbf{F}, \tag{11.14}$$

where ϵ_{nuc} is the nuclear energy generation rate per unit mass and $\mathbf{F}$ is the heat flux. In a steady state, where there are no velocities other than the rotational velocity u_ϕ, this reduces to

$$\text{div } \mathbf{F} = \rho \epsilon_{nuc}. \tag{11.15}$$

If energy transport is by radiation, then

$$\mathbf{F} = -\chi \nabla T, \tag{11.16}$$

where

$$\chi = \frac{4acT^3}{3\kappa\rho}, \tag{11.17}$$

where a is the radiation energy density constant and c is the speed of light. The opacity $\kappa(\rho, T)$ is a function of local fluid properties, and is also therefore just a function of Φ_e. This means that we can write

$$\mathbf{F} = f(\Phi_e)\nabla\Phi_e, \tag{11.18}$$

where the function $f(\Phi_e)$ is given by

$$f(\Phi_e) = -\frac{4acT^3}{3\kappa\rho}\frac{dT}{d\Phi_e}. \tag{11.19}$$

At the stellar surface, $p \to 0$ and $\rho \to 0$, and therefore $\Phi_e \to \Phi_s$ where Φ_s is the constant surface value of the effective potential. This means that on the stellar surface the radiative flux is proportional to the local surface gravity:

$$\mathbf{F} \propto \nabla\Phi_e \propto \mathbf{g}_e. \tag{11.20}$$

Since by definition the effective temperature T_e of the surface is given by the blackbody law, $F = \sigma T_e^4$, where $\sigma = ac/4$ is the Stefan–Boltzmann constant, we have finally that

$$T_e \propto g_e^{1/4}. \tag{11.21}$$

This is known as von Zeipel's law of gravity darkening. It implies that the temperature is not uniform over the surface of a rotating star, the equator being

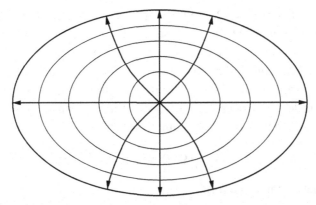

Fig. 11.1. Von Zeipel's theorem. In a star which rotates uniformly or on cylinders, the surfaces of constant temperature, pressure, density and effective gravity coincide. The T = const. surfaces are therefore more widely spaced at the equator than at the poles. Since heat flow is locally perpendicular to the T = const. surfaces, this leads to a lower surface flux there than at the poles (gravity darkening).

cooler and thus darker than the poles. The physical reason for this is that the surfaces of constant temperature within the star are further apart at the equator because of the lower gravity there. This reduces the temperature gradient there, and thus the radiation flux (eq. (11.16)), leading to a lower effective temperature at the photosphere (see Fig. 11.1).

However, when we try to satisfy the energy equation (eq. (11.15)) we run into problems. Using the expression for **F** given in eq. (11.18), we obtain

$$\text{div}\,\mathbf{F} = f'(\Phi_e)|\nabla\Phi_e|^2 + f(\Phi_e)\nabla^2\Phi_e. \tag{11.22}$$

Using Poisson's equation, $\nabla^2\Phi = 4\pi G\rho$, we find from the definition of Φ_e, eq. (11.11), that

$$\nabla^2\Phi_e = 4\pi G\rho - \frac{1}{R}\frac{d}{dR}(R^2\Omega^2). \tag{11.23}$$

Thus the energy equation can be written in the following form:

$$f'(\Phi_e)g^2 + f(\Phi_e)\left[4\pi G\rho - \frac{1}{R}\frac{d}{dR}(R^2\Omega^2)\right] = \rho\epsilon_{\text{nuc}}. \tag{11.24}$$

If the distribution of elements in the star is uniform (for example in a zero-age main-sequence star), then ϵ_{nuc} is a function of ρ and T, and so is just a function of Φ_e. Thus all the terms in eq. (11.24) are simply functions of Φ_e except for the factor g^2 and the second term in square brackets involving Ω.

If the rotation is uniform, $\Omega = \Omega_0 =$ const., then the rotational term is equal to $2\Omega_0^2$, which is a constant. In this case, for the equation to be satisfied for all values of Φ_e, i.e. everywhere in the star, we must demand that $f'(\Phi_e) = 0$, i.e. that $f(\Phi_e) =$ const. From the definition of f, we see that this is true only if $\kappa \propto T^3/\rho$. This is not true in general. If the rotation is not uniform, the same applies, and in addition, in order to make the term in square brackets depend only on Φ_e (so that we can equate this to the r.h.s.), we require further that

$$\frac{1}{R}\frac{d}{dR}(\Omega^2 R^2) = \text{const.} \tag{11.25}$$

This implies that Ω takes the form

$$\Omega^2 = C_1 + \frac{C_2}{R^2}, \tag{11.26}$$

where C_1 and C_2 are constants. However, then in order that Ω is finite throughout the star, we require $C_2 = 0$, and therefore that $\Omega =$ const. The energy equation then implies that ϵ_{nuc} take the following form:

$$\epsilon_{\text{nuc}} \propto \left(1 - \frac{\Omega^2}{2\pi G\rho}\right), \tag{11.27}$$

which is also not true in general.

To summarize: so far we have seen that for a uniformly rotating star, hydrostatic equilibrium implies that the stellar gas is barotropic (i.e. $p = p(\rho)$). The imposition of radiative equilibrium then leads to contradictions. Thus stars *either* have non-uniform rotation, *or* are not in radiative equilibrium.

11.3 Meridional circulation

The standard solution to this problem is to assume that the heat flux within the star is carried not only by conduction (radiative transfer) but is also transported by a steady circulation current in the meridional plane. The fluid flow within the star then has an additional component of the form

$$\mathbf{u}(R, z) = (u_R, 0, u_z). \tag{11.28}$$

Then the energy equation becomes

$$\rho T \mathbf{u} \cdot \nabla S = \text{div}(\chi \nabla T) + \rho \epsilon_{\text{nuc}}, \tag{11.29}$$

and we must also ensure mass conservation by taking

$$\text{div}(\rho \mathbf{u}) = 0. \tag{11.30}$$

If the star is not rotating too fast, so that the quantity $\eta = \Omega^2 R^3/GM$ is small, the standard procedure is to expand the equations in terms of η. There are now énough degrees of freedom to solve for the meridional circulation **u**, and we refer the reader in Section 11.5 to the various textbooks in which this procedure is carried out. Note that if there is no rotation, $\eta = 0$, there is no problem and we can take $u = 0$. Thus we expect the magnitude of u to be proportional to η. From eq. (11.29) we may estimate the velocity required to carry a heat flux η times that transported by conduction. To order of magnitude, we see that

$$\rho T u \frac{S}{R} \sim \eta \frac{1}{R}\frac{L}{R^2}. \tag{11.31}$$

Here L is the stellar luminosity, and we have approximated the thermally conducted flux as $\chi \nabla T \sim L/R^2$. Now the thermal energy content of the star is $\sim R^3 \rho TS$, and by the virial theorem for an equilibrium star we can equate this, to order of magnitude, to the gravitational self-energy $\sim GM^2/R$. Putting this together, we find

$$u \sim \eta \frac{LR^2}{GM} \sim \eta \frac{R}{t_{\mathrm{KH}}}. \tag{11.32}$$

Here t_{KH} is the time taken for a star to radiate away its thermal (or gravitational) energy . This Kelvin–Helmholtz timescale is around 30 million years for the Sun. The circulation timescale R/u is known as the Eddington–Sweet circulation time and is therefore given roughly by

$$t_{\mathrm{ES}} \sim t_{\mathrm{KH}}/\eta. \tag{11.33}$$

The solar rotation period is around 30 days, and thus for the Sun $\eta \sim 1/3000$. The resulting circulation timescale, $t_{\mathrm{ES}} \sim 10^{11}$ years, exceeds the age of the Universe. However for faster-rotating stars the circulation timescale can become comparable to the timescale for nuclear evolution. Meridional circulation can in principle move nuclear fuel around within such stars, and so can, again in principle, have a significant influence on the star's evolution.

11.3.1 The basic snag with meridional circulation

There is a fundamental problem with the standard procedure outlined above, which is well known to those in the field. We ran into a snag in trying to construct a model of a rotating star because we found that if we assumed that the star rotates uniformly, then it cannot also transport heat solely by conduction. The 'meridional circulation' solution to this is to assume the existence of a meridional flow which transports heat. The basic problem with this approach arises because the assumed meridional flow would also advect angular momentum. This then leads to the star no longer being in uniform rotation. Moreover, since we have assumed that the motions are steady, the postulated existence of a meridional flow requires a corresponding process neutralizing the advection of angular momentum by the flow, by transporting it back at the same rate as it is advected by the flow. For this to happen the star must become differentially rotating, so that $\Omega = \Omega(R, z)$, and there must be dissipative processes such as turbulence, magnetic field or convection which, when combined with the shear flow, can transport angular momentum in the right direction and at the right rate.

Garaud (2002) shows that this can be achieved for a simple molecular viscosity . For realistic stellar transport processes the details have yet to be worked out, and it has yet to be demonstrated that such a solution occurs in practice in real stars. Assuming, however, that something like this does actually occur, we may tentatively conclude that a rotating star is likely to be differentially rotating and that it will contain circulation currents of some sort. However, whether these currents are able to mix nuclear fuel, and so affect the stellar evolution, depends on unknown and uncertain dissipative processes within the star which determine the detailed, and as yet unknown, structure of the velocity field.

11.4 Rotation and magnetism

We have seen that the structure of a rotating star must in reality be quite complex because simple assumptions force unphysical constraints on the star. Something similar happens when we consider rotating stars with magnetic fields, as the following result shows. In a differentially rotating fluid, a steady magnetic field configuration exists only if the magnetic field lines lie along surfaces of constant angular velocity. This is Ferraro's law of isorotation.

To see this, suppose that the fluid rotates differentially with angular velocity $\Omega(R, z)$. Any magnetic field is advected by the fluid flow, and thus it is immediately apparent that the azimuthal field component, B_ϕ, is unaffected by the rotation. We may therefore simply consider a poloidal field of the form

$$\mathbf{B}(R, z) = (B_R, 0, B_z). \tag{11.34}$$

From the induction equation, we know that a steady field obeys

$$\mathrm{curl}(\mathbf{u} \wedge \mathbf{B}) = 0. \tag{11.35}$$

The ϕ-component of this equation is given by

$$\frac{\partial}{\partial z}(u_\phi B_z) + \frac{\partial}{\partial R}(u_\phi B_R) = 0. \tag{11.36}$$

Using the fact that $u_\phi = R\Omega$, this may be written in the form

$$\frac{\partial}{\partial z}(\Omega B_z) + \frac{1}{R}\frac{\partial}{\partial R}(R\Omega B_R) = 0. \tag{11.37}$$

We now note that this equation is div $(\mathbf{B}\Omega) = 0$, which when coupled with the Maxwell equation, div $\mathbf{B} = 0$, implies that

$$\mathbf{B} \cdot \nabla\Omega = 0. \tag{11.38}$$

Thus the magnetic field must lie at right angles to the gradient of the angular velocity; that is, the field lines must lie along the surfaces of constant angular velocity. On reflection, this is an obvious result. If a field line is such that different points on it have different angular velocities, then the field line will eventually twist up, creating an azimuthal magnetic field. Thus it cannot be in a steady state.

11.5 Further reading

A more detailed description of the theory of rotating stars can be found in the works of Tassoul (1978, 2000). A clear description of what is involved in the computation of meridional circulation in a simple idealized case is given by Garaud (2002).

11.6 Problems

11.6.1 Derive the equation of motion of an inviscid barotropic fluid subject to a conservative force field in a frame rotating with constant angular velocity $\mathbf{\Omega}$.

Show that in the rotating frame

$$\int_S (\boldsymbol{\omega} + 2\mathbf{\Omega}) \cdot \mathbf{dS} = \text{const.}, \tag{11.39}$$

where S is a surface spanning a closed contour Γ which moves with the fluid and where $\boldsymbol{\omega} = \text{curl}\,\mathbf{u}$, where $\mathbf{u}$ is the velocity in the rotating frame. (See Greenspan (1968, Chap. 1).)

11.6.2 An incompressible fluid moves steadily and *slowly* in a frame rotating with constant angular velocity $\mathbf{\Omega}$. Taking appropriate approximations, show that

$$(\mathbf{\Omega} \cdot \nabla)\mathbf{u} = 0. \tag{11.40}$$

What does this result imply physically? (This is the Taylor–Proudman theorem; see Greenspan (1968, Chap. 1).)

12

Circular shear flow

In this chapter we shall consider the stability of a differentially rotating fluid. Clearly this is directly relevant to important subjects in astrophysics, such as accretion disc stability and galactic structure. As before we shall consider some particular examples which serve to illustrate the relevant physical considerations.

12.1 Incompressible shear flow in a rigid cylinder

We consider first a rotating flow of an incompressible fluid within rigid cylindrical walls. Thus the density $\rho = \text{const.}$ and the unperturbed flow has a velocity field in cylindrical coordinates (R, ϕ, z) of the form

$$\mathbf{u}_0 = (0, V(R), 0), \tag{12.1}$$

where the azimuthal velocity $V(R)$ is related to the angular velocity $\Omega(R)$ by

$$V(R) = R\Omega(R). \tag{12.2}$$

For such a flow, with fluid velocity $\mathbf{u} = (u_R, u_\phi, u_z)$, the equations of motion are given by

$$\frac{\partial u_R}{\partial t} + \mathbf{u} \cdot (\nabla u_R) - \frac{u_\phi^2}{R} = -\frac{1}{\rho} \frac{\partial p}{\partial R}, \tag{12.3}$$

$$\frac{\partial u_\phi}{\partial t} + \mathbf{u} \cdot (\nabla u_\phi) + \frac{u_\phi u_R}{R} = -\frac{1}{R\rho} \frac{\partial p}{\partial \phi} \tag{12.4}$$

and

$$\frac{\partial u_z}{\partial t} + \mathbf{u} \cdot (\nabla u_z) = -\frac{1}{\rho} \frac{\partial p}{\partial z}. \tag{12.5}$$

158

In addition, we require the mass conservation equation, which for an incompressible fluid is just given by

$$\frac{\partial u_R}{\partial R} + \frac{u_R}{R} + \frac{1}{R}\frac{\partial u_\phi}{\partial \phi} + \frac{\partial u_z}{\partial z} = 0. \tag{12.6}$$

We note that since ρ is constant, the equilibrium pressure distribution is (from eq. (12.3)) simply given by

$$p(R) = \rho \int \frac{V^2}{R}\, dR. \tag{12.7}$$

We perturb this solution, so that the velocity field becomes

$$\mathbf{u} = (u'_R, V(R) + u'_\phi, u'_z). \tag{12.8}$$

To simplify notation we drop the primes from the first-order quantities. We must therefore note that u_ϕ is now the perturbation to the azimuthal velocity. We also note that there is no perturbation to the density, and for convenience we define the quantity

$$W = \frac{p'}{\rho}. \tag{12.9}$$

To first-order in small quantities, the equations of motion are given by

$$\frac{\partial u_R}{\partial t} + \frac{V}{R}\frac{\partial u_R}{\partial \phi} - \frac{2V u_\phi}{R} = -\frac{\partial W}{\partial R}, \tag{12.10}$$

$$\frac{\partial u_\phi}{\partial t} + \frac{V}{R}\frac{\partial u_\phi}{\partial \phi} + \left(\frac{V}{R} + \frac{dV}{dR}\right) u_R = -\frac{1}{R}\frac{\partial W}{\partial \phi} \tag{12.11}$$

and

$$\frac{\partial u_z}{\partial t} + \frac{V}{R}\frac{\partial u_z}{\partial \phi} = -\frac{\partial W}{\partial z}. \tag{12.12}$$

We note that the mass conservation equation, eq. (12.6), is already linearized.

Since the equilibrium configuration is independent of time, and of ϕ and z, we may now Fourier analyze in t, ϕ and z. Thus we assume that all the linear variables are of the form

$$p'(R, \phi, z, t) \rightarrow p'(R)\exp\{i(\omega t + m\phi + kz)\}. \tag{12.13}$$

In addition, we note that to keep p' a single-valued function of azimuth ϕ we require that m is an integer. It is also convenient to define a quantity $\sigma(R)$ which corresponds to the local Doppler-shifted frequency[†]

$$\sigma(R) = \omega + m\Omega(R). \tag{12.14}$$

[†] It is important to remember that σ is a function of radius R. In our experience, errors in algebra often occur because this fact is forgotten.

The linearized equations are now as follows:

$$i\sigma u_R - 2\Omega u_\phi = -\frac{dW}{dR},$$
(12.15)

$$i\sigma u_\phi + \left[\Omega + \frac{d}{dR}(R\Omega)\right]u_R = -\frac{imW}{R},$$
(12.16)

$$i\sigma u_z = -ikW$$
(12.17)

and

$$\frac{du_R}{dR} + \frac{u_R}{R} + \frac{imu_\phi}{R} + iku_z = 0.$$
(12.18)

We now write the equations in terms of the Lagrangian displacement $\boldsymbol{\xi}$. From eq. (4.24) in Chapter 4 we recall that if the unperturbed fluid has a flow field $\mathbf{u}_0$, then the Eulerian velocity perturbation $\mathbf{u}'$ is given in terms of $\boldsymbol{\xi}$ by

$$\mathbf{u}' = \frac{\partial \boldsymbol{\xi}}{\partial t} + \mathbf{u}_0 \cdot \nabla \boldsymbol{\xi} - \boldsymbol{\xi} \cdot \nabla \mathbf{u}_0.$$
(12.19)

In the case we are considering, with the unperturbed velocity field in the azimuthal direction and dependent only on radius R, we find that

$$u_R = i\sigma \xi_R$$
(12.20)

and

$$u_z = i\sigma \xi_z.$$
(12.21)

However, the azimuthal component takes a little more care, and we find that

$$u_\phi = i\sigma \xi_\phi - R\frac{d\Omega}{dR}\xi_R.$$
(12.22)

Thus the underlying shear flow implies that $u_\phi \neq i\sigma \xi_\phi$. The added complication to ξ_ϕ is necessary to ensure that when we write the mass conservation equation, eq. (12.18), in terms of $\boldsymbol{\xi}$ we obtain what is obviously the correct physical result:

$$\text{div}\,\boldsymbol{\xi} = 0.$$
(12.23)

We now use these relations to write the linearized equations in terms of $\boldsymbol{\xi}$. The equations are as follows:

$$(\sigma^2 - 2R\Omega\Omega')\xi_R + 2i\sigma \Omega\xi_\phi = \frac{dW}{dR},$$
(12.24)

$$\sigma^2\xi_\phi - 2i\sigma \Omega\xi_R = \frac{imW}{R},$$
(12.25)

$$\sigma^2\xi_z = ikW$$
(12.26)

and

$$\frac{d\xi_R}{dR} + \frac{\xi_R}{R} + \frac{im}{R}\xi_\phi + ik\xi_z = 0, \tag{12.27}$$

where we have used the notation $\Omega' = d\Omega/dR$.

We now eliminate ξ_ϕ and ξ_z from this set of equations, ending with two equations for the quantities W and ξ_R.

First, we multiply eq. (12.25) by im/R, multiply eq. (12.26) by ik and add, and then use eq. (12.27). Thus we obtain

$$\sigma^2 \left(\frac{d\xi_R}{dR} + \frac{\xi_R}{R} \right) - \frac{2m\Omega\sigma}{R}\xi_R = \left(\frac{m^2}{R^2} + k^2 \right) W. \tag{12.28}$$

Second, we eliminate ξ_ϕ between eqs. (12.24) and (12.25) to obtain

$$[\sigma^2 - \mathcal{R}(R)] = \frac{dW}{dR} + \frac{2m\Omega}{\sigma R}W. \tag{12.29}$$

Here $\mathcal{R}$ is the Rayleigh discriminant defined by

$$\mathcal{R}(R) = \frac{2\Omega}{R}\frac{d}{dR}(R^2\Omega), \tag{12.30}$$

which therefore depends on the radial gradient dj/dR of the specific angular momentum $j(R) = R^2\Omega$.

The various perturbation modes can now be found by solving eqs. (12.29) and (12.30), subject to the boundary condition that $\xi_R = 0$ at the inner and outer cylindrical boundaries $R = R_1, R_2$. Each modal solution will determine the eigenvalue $\omega = \sigma - m\Omega$, which determines the oscillation frquency, and the stability of the mode. For general non-axisymmetric modes ($m \neq 0$), it is not possible to give a simple local stability criterion. However, for axisymmetric modes the situation is different.

12.1.1 Axisymmetric perturbations: Rayleigh's criterion

For axisymmetric perturbations, $m = 0$ and therefore $\sigma = \omega$. Then eq. (12.28) becomes

$$\frac{1}{R}\frac{d}{dR}(R\xi_R) = \frac{k^2}{\omega^2}W \tag{12.31}$$

and eq. (12.29) becomes

$$[\omega^2 - \mathcal{R}] = \frac{dW}{dR}. \tag{12.32}$$

We can now simply eliminate W between these two equations to give

$$\frac{d}{dR}\left(\frac{1}{R}\frac{d}{dR}(R\xi_R) \right) - k^2\xi_R = -\frac{k^2\mathcal{R}(R)}{\omega^2}\xi_R. \tag{12.33}$$

For each value of k, given the boundary conditions $\xi_R = 0$ at $R = R_1, R_2$, this is now a Sturm–Liouville problem with eigenvalue $\lambda = k^2/\omega^2$. Multiplying both sides of the equation by $R\xi_R$ and integrating from R_1 to R_2, we find, after integrating by parts and using the boundary conditions, that

$$\frac{\omega^2}{k^2} = \frac{I_1}{I_2}, \tag{12.34}$$

where the integrals I_1 and I_2 are given by

$$I_1 = \int_{R_1}^{R_2} \mathcal{R}(R) R \xi_R^2 \, dR \tag{12.35}$$

and

$$I_2 = \int_{R_1}^{R_2} \left[\frac{1}{R} \left(\frac{d(R\xi_R)}{dR} \right)^2 + k^2 R \xi_R^2 \right] dR. \tag{12.36}$$

Since I_2 is positive definite, we see that the sign of ω^2 depends on the sign of I_1. This, in turn, depends on the sign of $\mathcal{R}(R)$. If $\mathcal{R} > 0$ at all radii, then $\omega^2 > 0$ and the flow is stable *to axisymmetric modes*. Conversely, if $\mathcal{R} < 0$ at some point in the flow, then we may choose a trial function $\xi(R)$ which makes $I_1 < 0$, which implies the existence of an unstable mode. This gives us *Rayleigh's criterion* for the stability of a circular shear flow, which states that the flow is stable *to axisymmetric disturbances* if and only if the specific angular momentum $j(R)$ increases outwards.

12.2 Axisymmetric stability of a compressible rotating flow

Of course, realistic astrophysical gases are generally compressible, so we need to extend the treatment of Section 12.1 to this case. In a compressible flow we must consider density variations. We have seen in Chapter 4 that in a horizontally stratified fluid in a vertical gravitational field the Schwarzschild stability criterion requires that the entropy of the fluid increases with height. In a rotating fluid there is an effective radial gravitational force caused by the centrifugal force, and so we might expect similar considerations to apply here. In addition, as we have seen, the axisymmetric stability of an incompressible rotating fluid to shear is decided by the Rayleigh criterion. Therefore in a rotating compressible fluid, we might expect stability to be governed by some combination of these two criteria. This is indeed the case, and this generalization is known as the Solberg–Høiland criterion; we derive this below.

We consider an axisymmetric fluid flow in a fixed axisymmetric gravitational potential $\Phi(R)$. The angular velocity is $\Omega(R)$ and the corresponding angular momentum is $j(R) = R^2\Omega$. Then the total energy of the unperturbed flow is given by

$$\mathcal{E} = \frac{1}{2}\int_V \frac{j^2}{R^2}\,dm + \int_V e\,dm + \int_V \Phi\,dm. \tag{12.37}$$

Here $dm = \rho\,dV$ is the mass of a fluid element and e is the internal energy. The integral is carried out over the volume V occupied by the fluid. Thus the three terms represent kinetic energy, thermal energy and gravitational potential energy, respectively.

12.2.1 Equilibrium

We now compute the change $\delta\mathcal{E}$ in the energy $\mathcal{E}$ when the flow is subject to an axially symmetric perturbation $\boldsymbol{\xi}(R)$. Since $\boldsymbol{\xi}$ is axisymmetric, the angular momentum j of each fluid element (or ring) is conserved. Thus the change in the kinetic energy occurs because the radius of each fluid element is changed; this is given by

$$\delta\frac{1}{2}\int_V \frac{j^2}{R^2}\,dm = -\int_V \frac{j^2}{R^3}\xi_R\,dm. \tag{12.38}$$

We assume that as fluid elements move they conserve their entropy, as we did when considering convective instability. From eq. (6.4) in Chapter 6 we recall that if entropy is conserved, i.e. $DS/Dt = 0$, then

$$\frac{De}{Dt} + \frac{p}{\rho}\,\text{div}\,\boldsymbol{u} = 0. \tag{12.39}$$

This implies that

$$\delta e = -\frac{p}{\rho}\,\text{div}\,\boldsymbol{\xi}. \tag{12.40}$$

Integrating by parts (recalling that $dV = dm/\rho$), we find that

$$\delta\int_V e\,dm = \int_V \frac{1}{\rho}\boldsymbol{\xi}\cdot\nabla p\,dm. \tag{12.41}$$

Since the potential Φ is fixed, we obtain

$$\delta\Phi = \boldsymbol{\xi}\cdot\nabla\Phi. \tag{12.42}$$

This implies that

$$\delta\int_V \Phi\,dm = \int_V \boldsymbol{\xi}\cdot\nabla\Phi\,dm. \tag{12.43}$$

Assembling our results we now obtain

$$\delta\mathcal{E} = \int_V \boldsymbol{\xi}\cdot\left(\nabla\Phi + \frac{1}{\rho}\nabla p - \frac{j^2}{R^3}\hat{\mathbf{R}}\right)dm. \tag{12.44}$$

For a fluid in equilibrium, we expect that $\delta\mathcal{E} = 0$ for all possible vector fields $\boldsymbol{\xi}$. This therefore implies that the expression in brackets in the integrand in eq. (12.44) must vanish everywhere. That is, the net force per unit mass acting on any fluid element is given by

$$\mathbf{F} = -\nabla\Phi - \frac{1}{\rho}\nabla p + \frac{j^2}{R^3}\hat{\mathbf{R}} = 0. \tag{12.45}$$

This is, of course, the time-independent momentum equation.

Another way of looking at this result is to note that the work done on a fluid element dm moving in a direction $\boldsymbol{\xi}$ is simply $\boldsymbol{\xi} \cdot \mathbf{F}\, dm$. Then if the fluid is in equilibrium the net work done must vanish, i.e.

$$\delta\mathcal{E} = -\int_V \boldsymbol{\xi} \cdot \mathbf{F}\, dm = 0. \tag{12.46}$$

This argument is a fluid version of d'Alembert's principle of virtual work used in statics.

12.2.2 Stability

To decide the stability of the configuration we need to consider the second-order perturbation to the energy. Using the equilibrium condition $\mathbf{F} = 0$, this is given by

$$\delta(\delta\mathcal{E}) = -\int_V \boldsymbol{\xi} \cdot \delta\mathbf{F}\, dm. \tag{12.47}$$

Recall that we are only allowing perturbations which conserve the specific angular momentum j and the entropy S of each fluid element. The configuration is stable to such a perturbation if $\delta^2\mathcal{E} > 0$.

To evaluate this we start with the integral expression for $\delta^2\mathcal{E}$, namely

$$\delta^2\mathcal{E} = \int_V \boldsymbol{\xi} \cdot \left[\delta\left\{\nabla\Phi + \frac{1}{\rho}p\right\} + \frac{3j^2}{R^4}\xi_R\hat{\mathbf{R}}\right] dm, \tag{12.48}$$

where we have used the fact that, for these perturbations,

$$\delta\left(\frac{1}{R^3}\right) = -\frac{3}{R^4}\xi_R. \tag{12.49}$$

For the first term in the square brackets in the integrand of eq. (12.48) we note that

$$\delta\left\{\nabla\Phi + \frac{1}{\rho}\nabla p\right\} = \left\{\nabla\Phi + \frac{1}{\rho}\nabla p\right\}' + \boldsymbol{\xi} \cdot \nabla\left\{\nabla\Phi + \frac{1}{\rho}\nabla p\right\}, \tag{12.50}$$

where the prime denotes Eulerian perturbation. Further, since the potential is fixed (and so has zero Eulerian perturbation), we have that

$$\left\{\nabla\Phi + \frac{1}{\rho}\nabla p\right\}' = \frac{1}{\rho}\nabla p' - \frac{\rho'}{\rho^2}\nabla\rho. \tag{12.51}$$

Using eq. (12.45) we see that the second term here can be written as follows:

$$\boldsymbol{\xi} \cdot \nabla \left\{ \nabla \Phi + \frac{1}{\rho} \nabla p \right\} = \boldsymbol{\xi} \cdot \nabla \left\{ \frac{j^2}{R^3} \hat{\mathbf{R}} \right\}. \tag{12.52}$$

We combine this with the second term in the square brackets in eq. (12.48) by noting that

$$\frac{3j^2}{R^4} \xi_R \hat{\mathbf{R}} + \boldsymbol{\xi} \cdot \nabla \left\{ \frac{j^2}{R^3} \hat{\mathbf{R}} \right\} = (\boldsymbol{\xi} \cdot \nabla j^2) \frac{\hat{\mathbf{R}}}{R^3}. \tag{12.53}$$

Putting this all together, we now find that for axisymmetric perturbations

$$\delta^2 \mathcal{E} = - \int_V \boldsymbol{\xi} \cdot (\mathcal{L} \boldsymbol{\xi}) dm, \tag{12.54}$$

where the linear operator $\mathcal{L}$ is given by

$$\mathcal{L} \boldsymbol{\xi} = -\frac{\nabla p'}{\rho} + \frac{\rho'}{\rho^2} \nabla \rho - (\boldsymbol{\xi} \cdot \nabla j^2) \frac{\hat{\mathbf{R}}}{R^3}. \tag{12.55}$$

We note that, apart from the rotational terms, this is essentially the same result as we obtained in Chapter 10. We now follow the analysis of that chapter, where we combined the continuity equation,

$$\rho' = -\text{div}(\rho \boldsymbol{\xi}), \tag{12.56}$$

and the adiabaticity of the perturbation in the form

$$p' = -\gamma p \, \text{div} \, \boldsymbol{\xi} - \boldsymbol{\xi} \cdot \nabla p \tag{12.57}$$

to conclude that, in this case,

$$\delta^2 \mathcal{E} = \int_V \xi_i \mathcal{M}_{ij} \xi_j \, dm + \int_V \left[\frac{(p')^2}{\gamma p \rho} \right] dm, \tag{12.58}$$

where the second-order tensor $\mathcal{M}_{ij}$ is defined as follows:

$$\mathcal{M}_{ij} = \left[\frac{1}{\rho} \nabla p \right]_i \left[\frac{1}{\rho} \nabla \rho - \frac{1}{\gamma p} \nabla p \right]_j + \frac{1}{R^3} (\nabla R)_i (\nabla j^2)_j. \tag{12.59}$$

12.2.3 The Solberg–Høiland criterion

We now recognize that for such axisymmetric perturbations the effect of rotation is to add into eq. (12.58) a term involving ∇j^2. Thus, as we expected, the stability of the configuration must involve some combination of the Schwarzschild stability criterion to convection we derived in Chapter 4 and the Rayleigh criterion for rotation we derived in Section 12.1.1. We recall that for stability we require $\delta^2 \mathcal{E} > 0$. The second term on the r.h.s of eq. (12.58) is positive definite. For stability considerations we need therefore only consider those perturbations for which this term vanishes. Thus we need only concentrate our analysis on the first term on the r.h.s.

We first show that $\mathcal{M}$ is a symmetric tensor. We define the following vectors:

$$\mathbf{A} = \frac{1}{\gamma p}\nabla p - \frac{1}{\rho}\nabla\rho, \tag{12.60}$$

$$\mathbf{A}' = -\frac{1}{\rho}\nabla\rho, \tag{12.61}$$

$$\mathbf{B} = \frac{1}{R^3}\nabla j^2 \tag{12.62}$$

and

$$\mathbf{B}' = \nabla R = \hat{\mathbf{R}}, \tag{12.63}$$

so that we may write

$$\mathcal{M}_{ij} = A_i A_j' + B_i B_j'. \tag{12.64}$$

Taking the curl of the equilibrium equation, eq. (12.45), we obtain

$$\nabla\left(\frac{1}{\rho}\right)\wedge\nabla p = \frac{1}{R^3}\nabla j^2 \wedge \hat{\mathbf{R}}, \tag{12.65}$$

which implies that

$$\mathbf{A}\wedge\mathbf{A}' + \mathbf{B}\wedge\mathbf{B}' = 0. \tag{12.66}$$

In suffix notation this implies that

$$\epsilon_{ijk}\mathcal{M}_{jk} = 0, \tag{12.67}$$

and therefore that $\mathcal{M}$ is symmetric.

The condition for a real symmetric second-rank tensor to give a positive definite expression when contracted twice with a vector, as in eq. (12.58), is simply that $\mathrm{tr}(\mathcal{M}) > 0$ and $\det(\mathcal{M}) > 0$. This gives us the Solberg–Høiland criterion that a fluid configuration of the kind we are considering is stable to axisymmetric, adiabatic perturbations if and only if

$$\frac{1}{R^3}\frac{\partial j^2}{\partial R} + \frac{1}{c_p}(-\mathbf{g}\cdot\nabla S) > 0 \tag{12.68}$$

and

$$-g_z\left(\frac{\partial j^2}{\partial R}\frac{\partial S}{\partial z} - \frac{\partial j^2}{\partial z}\frac{\partial S}{\partial R}\right) > 0. \tag{12.69}$$

Here we have defined gravity **g** as follows:

$$\mathbf{g} = \frac{1}{\rho}\nabla p = -\mathbf{A}', \tag{12.70}$$

and we have noted that

$$\mathbf{A} = \frac{\nabla S}{c_p}. \tag{12.71}$$

Thus, as expected, we obtain a blend of the Rayleigh and Schwarzschild stability criteria, as we can see by successively assuming that S and j^2 are spatially constant.

12.3 Circular shear flow with a magnetic field

We now consider the same (incompressible) flow as we did before in Section 12.1 except that we now add a constant uniform magnetic field given by

$$\mathbf{B}_0 = (0, 0, B) \tag{12.72}$$

to the unperturbed configuration. Note that a uniform field has no effect on the equilibrium. When the fluid has been perturbed, the field has the form

$$\mathbf{B} = \mathbf{B}_0 + \mathbf{b}, \tag{12.73}$$

where the field perturbation has components

$$\mathbf{b} = (b_R, b_\phi, b_z). \tag{12.74}$$

As before, the perturbed velocity has the form

$$\mathbf{u} = (u_R, V(R) + u_\phi, u_z) \tag{12.75}$$

and we define $\Omega(R) = V/R$ and $W = p'/\rho$.

As before, we consider only axisymmetric perturbations, and so take $\partial/\partial\phi = 0$ throughout. Then, taking account of the additional terms arising from the magnetic field, the linearized equations of motion are given by

$$\frac{\partial u_R}{\partial t} - 2\Omega u_\phi - B\frac{\partial b_R}{\partial z} = -\frac{\partial W}{\partial R}, \tag{12.76}$$

$$\frac{\partial u_\phi}{\partial t} + \left(\frac{dV}{dR} + \frac{V}{R}\right)u_R - B\frac{\partial b_\phi}{\partial z} = 0 \tag{12.77}$$

and

$$\frac{\partial u_z}{\partial t} - B\frac{\partial b_z}{\partial z} = -\frac{\partial W}{\partial z}. \tag{12.78}$$

In addition, we have now the linearized version of the induction equation describing the magnetic field evolution. The three components of this are given by

$$\frac{\partial b_R}{\partial t} - B\frac{\partial u_R}{\partial z} = 0, \tag{12.79}$$

$$\frac{\partial b_\phi}{\partial t} - B\frac{\partial u_\phi}{\partial z} - \left(\frac{dV}{dR} - \frac{V}{R}\right)b_R = 0 \tag{12.80}$$

and

$$\frac{\partial b_z}{\partial t} - B\frac{\partial u_z}{\partial z} = 0. \tag{12.81}$$

In addition, for an incompressible fluid we have

$$\frac{\partial u_R}{\partial R} + \frac{u_R}{R} + \frac{\partial u_z}{\partial z} = 0, \tag{12.82}$$

and since **B** is solenoidal we have

$$\frac{\partial b_R}{\partial R} + \frac{b_R}{R} + \frac{\partial b_z}{\partial z} = 0. \tag{12.83}$$

As before, we now Fourier analyze, but now for axisymmetry with $m = 0$. Thus all variables have a factor of the form $\exp\{i(\omega t + kz)\}$.

The linearized equations of motion become

$$i\omega u_R - 2\Omega u_\phi + ikBb_R = -\frac{dW}{dR}, \tag{12.84}$$

$$i\omega u_\phi + \left(\frac{dV}{dR} + \frac{V}{R}\right) - ikBb_\phi = 0 \tag{12.85}$$

and

$$i\omega u_z - ikBb_z = -ikW. \tag{12.86}$$

The components of the linearized induction equation are as follows:

$$i\omega b_R = ikBu_R, \tag{12.87}$$

$$i\omega b_\phi = ikBu_\phi + \left(\frac{dV}{dR} - \frac{V}{R}\right)b_R \tag{12.88}$$

and

$$i\omega b_z = ikBu_z. \tag{12.89}$$

We recall the relation between the linearized velocity **u** and the Lagrangian perturbation $\boldsymbol{\xi}$ which we derived in Section 12.1:

$$u_R = i\omega\xi_R, \tag{12.90}$$

$$u_\phi = i\omega\xi_\phi - R\frac{d\Omega}{dR}\xi_R \tag{12.91}$$

and

$$u_z = i\omega\xi_z. \tag{12.92}$$

From this we note, from a simple comparison of the equations, that

$$\mathbf{b} = ikB\boldsymbol{\xi}. \tag{12.93}$$

This is the linearized equivalent of the statement that magnetic field lines are dragged along with the fluid.

Using these results, we can replace (u_R, u_ϕ, u_z) and (b_R, b_ϕ, b_z) in the equations of motion by terms involving only (ξ_R, ξ_ϕ, ξ_z). We thus obtain the following three equations:

$$(\omega^2 - 2R\Omega\Omega' - \Omega_A^2)\xi_R + 2i\omega\Omega\xi_\phi = \frac{dW}{dR}, \tag{12.94}$$

$$(\omega^2 - \Omega_A^2)\xi_\phi - 2i\omega\Omega\xi_R = 0 \tag{12.95}$$

and

$$(\omega^2 - \Omega_A^2)\xi_z = ikW, \tag{12.96}$$

where we define the Alfvén frequency Ω_A as follows:

$$\Omega_A^2 = k^2 V_A^2, \tag{12.97}$$

where $V_A = \sqrt{B^2/\rho}$ is the Alfvén speed.

In addition to these, we have the continuity equation for an incompressible fluid:

$$\frac{d\xi_R}{dR} + \frac{\xi_R}{R} + ik\xi_z = 0. \tag{12.98}$$

Eliminating ξ_z between eqs. (12.96) and (12.98), we obtain

$$(\omega^2 - \Omega_A^2)\left(\frac{d\xi_R}{dR} + \frac{\xi}{R}\right) = k^2 W, \tag{12.99}$$

and we eliminate ξ_ϕ between eqs. (12.94) and (12.95) to obtain

$$\left(\omega^2 - \Omega_A^2 - 2R\Omega\Omega' - \frac{4\Omega^2\omega^2}{\omega^2 - \Omega_A^2}\right)\xi_R = \frac{dW}{dR}. \tag{12.100}$$

Using the definition of the Rayleigh discriminant,

$$\mathcal{R}(R) = \frac{2\Omega}{R}\frac{d}{dR}(R^2\Omega), \tag{12.101}$$

eq. (12.100) can be written as follows:

$$\left(\omega^2 - \Omega_A^2 - \mathcal{R}(R) - \frac{4\Omega^2\Omega_A^2}{\omega^2 - \Omega_A^2}\right)\xi_R = \frac{dW}{dR}. \tag{12.102}$$

12.3.1 Local analysis

We could now, in principle, for a particular shear flow $V(R)$, solve the two coupled
first-order differential equations (eqs. (12.99) and (12.102)) for W and ξ_R, and in
doing so obtain the eigenvalue ω which determines the stability. However, as before,
it is more instructive to look at modes with a short radial wavelength and so obtain a
local dispersion relation. Thus we assume that the spatial variation of the variables
is $\propto \exp\{i(k_R R + k_z z)\}$. Note that for clarity we have now replaced k by k_z. We also
assume that the radial wavenumber is large (i.e. the radial wavelength is small),
so that $k_R R \gg 1$. The two differential equations now become algebraic equations.
Thus eq. (12.99) becomes

$$(\omega^2 - \Omega_A^2)(ik_R R) = k_z^2 W, \tag{12.103}$$

and eq. (12.102) becomes

$$\left(\omega^2 - \Omega_A^2 - \mathcal{R}(R) - \frac{4\Omega^2 \Omega_A^2}{\omega^2 - \Omega_A^2}\right)\xi_R = ik_R W. \tag{12.104}$$

We can now eliminate W and ξ_R between these two equations and so obtain the
local dispersion relation:

$$I_1(\omega^2 - \Omega_A^2)^2 - I_2(\omega^2 - \Omega_A^2) - I_3 = 0, \tag{12.105}$$

where

$$I_1 = 1 + \frac{k_R^2}{k_z^2} > 0, \tag{12.106}$$

$$I_2 = \mathcal{R}(R) = 4\Omega^2 + R\frac{d\Omega^2}{dR} \tag{12.107}$$

and

$$I_3 = 4\Omega^2 \Omega_A^2 > 0. \tag{12.108}$$

The solutions of eq. (12.105) are as follows:

$$\omega^2 = \Omega_A^2 + \frac{I_2 \pm \sqrt{I_2^2 + 4I_1 I_3}}{2I_1}. \tag{12.109}$$

The solution with the $+$ sign always has $\omega^2 > 0$ and so is stable. Thus, since $I_1 > 0$,
for stability we require

$$2I_1 \Omega_A^2 + I_2 - \sqrt{I_2^2 + 4I_1 I_3} > 0. \tag{12.110}$$

After some manipulation to get rid of the square root, we find that this is true if and
only if

$$I_1 \Omega^4 > I_3 - I_2 \Omega_A^2. \tag{12.111}$$

A little further manipulation using eq. (12.107) to replace I_2 shows that stability to axisymmetric perturbations occurs if and only if

$$\frac{d\Omega^2}{dR} > 0. \tag{12.112}$$

12.3.2 The Balbus–Hawley instability

We see that the stability criterion for a rotating shear flow with a magnetic field is completely different from the criterion for one without. For a purely hydrodynamic flow, instability (to axisymmetric perturbations) occurs when the angular momentum of the fluid decreases outwards. In that case axisymmetric perturbations conserve the angular momentum of fluid elements and so can only release energy (and so drive an instability) when the underlying angular momentum profile is appropriately arranged. In contrast, the addition of a magnetic field implies that even though the perturbations are still axisymmetric, the angular momentum of each fluid element is no longer conserved because the magnetic field can apply stresses which move it about. Under these circumstances, instability to axisymmetic perturbations occurs when the angular *velocity* decreases outwards. This 'magnetorotational' instability (MRI) has been in the literature for some time (for example the derivation given here is based on that in the book by Chandrasekhar (1961)). However, its relevance to astrophysical phenomena, and especially to accretion discs, was not realized until the work of Balbus and Hawley in the early 1990s. They showed that the MRI provides a convincing basis for the long-sought mechanism driving outward transport of angular momentum, and thus inward transport of mass, in accretion discs.

One particularly interesting aspect of this instability is that although its occurrence depends on the presence of a magnetic field, the stability criterion does not depend on the size of the field. Thus any magnetic field, *however small*, gives rise to the instability! Physically all the field does is to allow fluid elements to exchange angular momentum. This implies that the Rayleigh criterion, which relies on fluid elements retaining their angular momenta, can be circumvented, enabling the shear flow energy to be liberated more easily. Even so, it seems strange at first glance that the stability criterion in the limit $B \to 0$ is quite different from the stability criterion for the case $B = 0$. This is an example of a singular limit found in many areas of physics. For example, a viscous fluid in the limit of vanishing viscosity behaves differently from an inviscid fluid . In his article on singular limits, Michael Berry (2002) writes:

Biting into an apple and finding a maggot is unpleasant enough, but finding half a maggot is worse. Discovering one third of a maggot is more distressing still. The less you find the more you might have eaten. Extrapolating to the limit, an encounter with no maggot at all should be the ultimate bad apple experience. This remorseless logic fails, however, because the limit is singular. A very small maggot fraction ($f \ll 1$) is qualitatively different from no maggot ($f = 0$).

12.4 Circular shear flow with self-gravity

In Chapter 9 we discussed the stability of fluid configurations subject to self-gravity. While doing so we noted that although we did find stability criteria, in each case the underlying fluid configuration was taken to be static but was not in fact in equilibrium. Further, the timescale for the development of the instabilities is typically $\sim (G\rho)^{-1/2}$, which is of course the same as the timescale for the evolution of the background fluid. However, now that we are in a position to add rotation, we are able to set up an initial configuration which is in dynamical equilibrium.

12.4.1 Rotating thin disc

We consider an initially axisymmetric, infinitesimally thin disc lying in the z-plane, with surface density $\Sigma(R)$. Thus the density distribution is given by

$$\rho(R, z) = \Sigma(R)\delta(z). \tag{12.113}$$

The (two-dimensional) pressure in the disc, $P(R)$, just depends on radius. The unperturbed velocity is an axisymmetric shear flow of the form

$$\mathbf{u}_0 = (0, R\Omega(R), 0), \tag{12.114}$$

and therefore the equilibrium condition just represents the balance between centrifugal force, the (radial) pressure gradient and the radial gradient of the gravitational force:

$$R\Omega^2 = -\frac{\nabla P}{\Sigma} - \nabla_\perp \Phi, \tag{12.115}$$

where Φ is the unperturbed gravitational potential. To keep the analysis simple, we shall assume that initially both the density and the pressure are uniform. Thus $\Sigma = $ const. and $\nabla P = 0$. Then the unperturbed angular velocity is given in terms of the gravitational potential by

$$\Omega^2 = -\frac{1}{R}\nabla_\perp \Phi. \tag{12.116}$$

12.4.1.1 Axisymmetric perturbations – the Toomre criterion

We consider axisymmetric perturbations within the disc plane. The perturbed velocity field is given by

$$\mathbf{u} = (u_R, R\Omega + u_\phi, 0). \tag{12.117}$$

We also Fourier analyze in time, so that all linear quantities vary as $\propto \exp(i\omega t)$.

The linearized equations of motion (in the disc plane only) are then given by

$$i\omega u_R - 2\Omega u_\phi = -\frac{1}{\Sigma}\frac{dP'}{dR} - \frac{d\Phi'}{dR} \tag{12.118}$$

and

$$i\omega u_\phi + \left[\Omega + \frac{d}{dR}(R\Omega)\right]u_R = 0, \tag{12.119}$$

where P' is the perturbed pressure and Φ' is the perturbed gravitational potential. The linearized continuity equation is given by

$$i\omega\Sigma' + \Sigma\left(\frac{du_R}{dR} + \frac{u_R}{R}\right) = 0, \tag{12.120}$$

where Σ' is the perturbed surface density. Linearization of Poisson's equation yields

$$\nabla^2\Phi' = 4\pi G\Sigma'\delta(z). \tag{12.121}$$

To simplify the analysis further, we now consider perturbations with short radial wavelength. This means that we can replace radial derivatives of perturbed quantities by the multiplicative factor ik, where k is the radial wavenumber, and we assume that $kR \gg 1$. In this approximation we saw in Chapter 9 that the perturbed gravitational potential in the $z = 0$ plane is given by

$$\Phi'(z = 0) = -\frac{2\pi G\Sigma'}{|k|}. \tag{12.122}$$

Using this, and defining the two-dimensional sound speed C_s in the disc by

$$C_s^2 = \frac{dP}{d\Sigma}, \tag{12.123}$$

we can combine eqs. (12.118) and (12.119) to eliminate u_ϕ and obtain

$$-\omega^2 u_R + 2\Omega\left[\Omega + \frac{d}{dR}(R\Omega)\right]u_R = -ikC_s^2\frac{i\omega\Sigma'}{\Sigma} - ik\left(-\frac{2\pi G}{|k|}(i\omega\Sigma')\right). \tag{12.124}$$

We now use eq. (12.120) to eliminate Σ' and find a linear equation for u_R:

$$-\omega^2 u_R + \left[4\Omega^2 + 2R\Omega\frac{d\Omega}{dR}\right]u_R = -k^2C_s^2 u_R + 2\pi G|k|\Sigma u_R. \tag{12.125}$$

We note that the term in square brackets is simply the Rayleigh discriminant $\mathcal{R}$, which we can write in terms of the epicyclic frequency κ as follows:

$$\mathcal{R} = \kappa^2(R). \tag{12.126}$$

Thus, on cancelling u_R from the equation we obtain a local dispersion relation for axisymmetric radial waves:

$$\omega^2 = \kappa^2 - 2\pi G|k|\Sigma + k^2C_s^2. \tag{12.127}$$

If there is no self-gravity then we have the usual dispersion relation for sound waves ($\omega^2 = k^2 C_s^2$) modified by rotation. The introduction of self-gravity introduces the possibility of instability. For large wavenumber (small wavelength) we still have stable sound waves, but they are now modified also by the self-gravity term. In addition, for small wavenumber (long wavelength) we also have stability, as $\omega \sim \kappa$. Thus instability can only occur, if it does at all, for some intermediate range of wavelengths. Instability requires $\omega^2 < 0$, and inspection of eq. (12.127) shows that this requires

$$Q = \frac{\kappa C_s}{\pi G \Sigma} < 1. \tag{12.128}$$

This is the Toomre criterion. When this is satisfied, there is a range of values of k for which instability occurs ($\omega^2 < 0$).

The criterion can be understood roughly in the following manner. Consider a patch of the disc of size λ. The mass contained within such a disturbance is $\Delta M \sim \Sigma \lambda^2$. Thus the timescale on which such a patch would collapse under gravity is $\tau_G \sim (\lambda^3 / G \Delta M)^{1/2} \sim (\lambda / G \Sigma)^{1/2}$. This patch can be stabilized against collapse by pressure and by shear (or angular momentum). The timescale on which shear operates is $\sim 1/\kappa$. Thus for collapse to be possible we require $\tau_G \leq 1/\kappa$, that is

$$\lambda \leq \frac{G \Sigma}{\kappa^2}. \tag{12.129}$$

The timescale on which pressure stabilizes is just the time taken for sound to cross the patch, that is $\tau_s \sim \lambda / C_s$. For collapse to be possible we also require $\tau_G \leq \tau_s$, that is

$$\lambda \geq \frac{C_s^2}{G \Sigma}. \tag{12.130}$$

For instability we require that both these inequalities are satisfied, and we therefore require that

$$\frac{G \Sigma}{\kappa^2} \leq \frac{C_s^2}{G \Sigma}. \tag{12.131}$$

This, roughly, is the Toomre criterion.

12.4.1.2 Non-axisymmetric disturbances – spiral arms

In astronomy, thin discs of gas in axisymmetric potentials are found in many situations, of which obvious examples are Saturn's rings and late type galaxies. Late type galaxies are also known as spiral galaxies because the gas (and newly born stars) in them display a spiral pattern. These spirals come about through

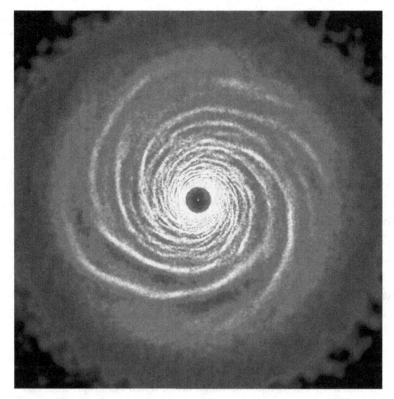

Fig. 12.1. Surface density given by a numerical calculation of a self-gravitating plane disc. The gravitational potential is that of a point mass at the centre of the disc and the disc mass is one-tenth of that of the central object. The mean value of Q in the disc is $\langle Q \rangle = 1.0$–1.2, so it is marginally stable to axisymmetric perturbations. The disc develops non-axisymmetric spiral self-gravitational instabilities. Courtesy of Giuseppe Lodato.

self-gravitational instability (see Fig. 12.1). This tells us that the axisymmetric instability we have considered so far, which would give rise to rings rather than spirals, is not commonly found. Why is this?

We found in Section 12.3 that the addition of a small magnetic field to a shear flow made it easier to tap the energy available in the shear, and so made it easier to drive an instability. This came about because in a purely axisymmetric non-magnetic perturbation the angular momentum of individual fluid elements is conserved. Then the relevant criterion was the Rayleigh criterion. However, the addition of a magnetic field, even arbitrarily small, releases the system from this constraint and so enables it to drive instability more easily (the stability criterion becomes that of Balbus and Hawley). Similar considerations apply here: non-axisymmetric perturbations break the angular momentum constraint for individual fluid elements in an exactly analogous way to the magnetic perturbations in the Balbus–Hawley instability. If the perturbation is non-axisymmetric, the linearized equation for u_ϕ has terms of the form $\partial P/\partial\phi$ and $\partial\Phi/\partial\phi$ on the r.h.s. These represent forces acting in the azimuthal direction, and therefore mean that the angular momentum of individual fluid elements is no longer constant. Thus, for non-axisymmetric perturbations the shear term (κ^2) is no longer such a stabilizing influence, and we expect it to be easier to drive instability.

12.5 Further reading

The underlying concepts required for the consideration of the stability of rotational shear flows are set out in Chandrasekhar (1961, Chap. VII). The derivation of the Solberg–Høiland criterion sented here is based on that given by Tassoul (2000, Chap. 3). Considerations of the effect of a magnetic field on rotational shear flow are given in Chandrasekhar (1961, Chap. IX). The relevance of magnetically induced instability of rotational shear flow in the astrophysical context is reviewed by Balbus & Hawley (1998). The original description of the stability of a self-gravitating disc, albeit for a system of stars, is given by Toomre (1964).

12.6 Problems

12.6.1 Show that the vorticity equation for an inviscid non-barotropic fluid can be written as follows:

$$\frac{D}{Dt}\left(\frac{\boldsymbol{\omega}}{\rho}\right) = \left(\frac{\boldsymbol{\omega}}{\rho}\right)\cdot\nabla\mathbf{u} - \frac{1}{\rho}\nabla\left(\frac{1}{\rho}\right)\wedge\nabla p. \tag{12.132}$$

Note that this implies that there can be a source of vorticity in a non-barotropic fluid. Show that the gradient of the enthalpy $h = e + p/\rho$ is given by

$$\nabla h = T\nabla S + \frac{1}{\rho}\nabla p, \tag{12.133}$$

where S is the entropy. Deduce that

$$\nabla T \wedge \nabla S = -\nabla\left(\frac{1}{\rho}\right)\wedge\nabla p. \tag{12.134}$$

Show also that for any scalar quantity Q

$$\frac{D}{Dt}\left(\frac{\boldsymbol{\omega}}{\rho}\cdot\nabla Q\right) = \frac{D}{Dt}\left(\frac{\boldsymbol{\omega}}{\rho}\right)\cdot\nabla Q + \frac{\boldsymbol{\omega}}{\rho}\cdot\frac{D}{Dt}[\nabla Q]. \tag{12.135}$$

Use the above results to show that the entropy obeys the following equation:

$$\frac{D}{Dt}\left(\frac{\boldsymbol{\omega}}{\rho}\cdot\nabla S\right) = \left(\frac{\boldsymbol{\omega}}{\rho}\right)\cdot\nabla\left[\frac{DS}{Dt}\right]. \tag{12.136}$$

(See Tassoul (1978, Chap. 3).)

12.6.2 Show that in axisymmetric motion, the specific angular momentum $L = Ru_\phi = R^2\Omega$ of a fluid element remains constant as we follow its motion. Show that the radial (R) and axial (z) equations of motion can be written in a form replacing any terms in u_ϕ with a force L^2/R^3 acting in the radial direction.

An incompressible fluid with uniform density ρ is in cylindrical shear flow, with angular velocity $\Omega(R)$. Consider the exchange of two elementary annuli (rings) of fluid of equal heights and mass, caused by axisymmetric fluid motions. The rings are at R_1 and R_2 and have small radial extents dR_1 and dR_2 and masses $dm = 2\pi\rho R_1 dR_1 = 2\pi\rho R_2 dR_2$. Calculate the change in energy caused by the exchange of the rings and show how this is related to the Rayleigh stability criterion.

13

Modes in rotating stars

We have seen in earlier chapters that in a non-rotating star there are basically two types of wave modes possible. The p-modes resemble sound waves as the restoring force in the oscillation comes from pressure. The g-modes resemble surface water waves as their restoring force comes from gravity. In this chapter we consider the new effects which can arise if the star is rotating. To do this we need a more sophisticated model for a star. Previously we took our 'star' as either a horizontally stratified atmosphere or as a gas in a one-dimensional box. To consider rotational effects we need an axisymmetric base state. In line with our philosophy of keeping things as simple as possible, without jettisoning any essential physics, we take as our model star a circular cylinder of incompressible fluid. This eliminates the consideration of p-modes.

13.1 The non-rotating 'star'

We start by considering the wave modes in the non-rotating 'star'. In equilibrium this is a cylinder of fluid of radius R_0, with constant density ρ and pressure $p(R)$, sitting in a gravitational potential $\Phi(R)$. The gravitational potential may be imposed from outside: it may result entirely from self-gravity of the cylinder, or it may be a mixture of the two. In equilibrium the velocity field is zero and the equation of hydrostatic equilibrium yields

$$-\frac{1}{\rho}\frac{\mathrm{d}p}{\mathrm{d}R} - \frac{\mathrm{d}\Phi}{\mathrm{d}R} = 0. \tag{13.1}$$

Initially we consider waves with no z-dependence; that is, waves for which $u_z = 0$ and $\partial/\partial z = 0$. Thus all we expect in this situation is waves travelling around the curved surface of the cylinder in the azimuthal direction. In this case the linearized equations of motion are given by

$$\frac{\partial u_R}{\partial t} = -\frac{1}{\rho}\frac{\partial p'}{\partial R} - \frac{\partial \Phi'}{\partial R} - \frac{\rho'}{\rho}\frac{d\Phi}{dR} \qquad (13.2)$$

and

$$\frac{\partial u_\phi}{\partial t} = -\frac{1}{R\rho}\frac{\partial p'}{\partial \phi} - \frac{1}{R}\frac{\partial \Phi'}{\partial \phi}. \qquad (13.3)$$

We emphasize that although the fluid is incompressible, so that $\delta\rho = 0$, the Eulerian density perturbation ρ' is non-zero because the fluid surface is distorted by the perturbation. Thus if the radial component of the Lagrangian perturbation is ξ_R, the Eulerian density perturbation is given by

$$\rho' = \rho\xi_R\delta(R - R_0). \qquad (13.4)$$

The linearized continuity equation is given by

$$\frac{1}{R}\frac{\partial}{\partial R}(Ru_R) + \frac{1}{R}\frac{\partial u_\phi}{\partial \phi} = 0, \qquad (13.5)$$

and the linearized version of Poisson's equation is given by

$$\nabla^2\Phi' = 4\pi G\rho'. \qquad (13.6)$$

Using the symmetry of the unperturbed configuration, we can now Fourier analyze in the usual way by taking all quantities to vary as $\propto \exp\{i(\omega t+m\phi+kz)\}$. In this case we have assumed no z-dependence, so that $k = 0$. Making this substitution, and defining the quantity W by

$$W = \frac{p'}{\rho} + \Phi', \qquad (13.7)$$

the above equations become

$$i\omega u_R = -\frac{dW}{dR} - \frac{\rho'}{\rho}\frac{d\Phi}{dR}, \qquad (13.8)$$

$$i\omega u_\phi = \frac{imW}{R}, \qquad (13.9)$$

$$\frac{1}{R}\frac{d}{dR}(Ru_R) + \frac{imu_\phi}{R} = 0 \qquad (13.10)$$

and

$$\frac{1}{R}\frac{d}{dR}\left(R\frac{d\Phi'}{dR}\right) - \frac{m^2}{R^2}\Phi' = 4\pi G\rho\xi_R\delta(R - R_0). \qquad (13.11)$$

We eliminate u_ϕ between eqs. (13.9) and (13.10) to obtain

$$\frac{i\omega}{R}\frac{d}{dR}(Ru_R) = \frac{m^2W}{R^2}. \qquad (13.12)$$

Then using eq. (13.8) everywhere *except* at $R = R_0$ (i.e. wherever $\rho' = 0$)

$$i\omega u_R = -\frac{\mathrm{d}W}{\mathrm{d}R}, \tag{13.13}$$

we obtain an equation for W valid everywhere *except* at $R = R_0$:

$$\frac{1}{R}\frac{\mathrm{d}}{\mathrm{d}R}\left(R\frac{\mathrm{d}W}{\mathrm{d}R}\right) - \frac{m^2 W}{R^2} = 0. \tag{13.14}$$

Inside the star, $R < R_0$, the solution that is finite at the origin is given by

$$W(R) \propto R^m, \tag{13.15}$$

and we assume that $m \neq 0$.

Similarly, the solution of eq. (13.11) is straightforward except at $R = R_0$. Choosing the solutions which are finite at the origin and at infinity we find (for $m \neq 0$) that

$$\Phi' = \begin{cases} \Phi'_0 (R/R_0)^m, & R < R_0, \\ \Phi'_0 (R/R_0)^{-m}, & R > R_0. \end{cases} \tag{13.16}$$

Here Φ'_0 is the value of Φ' at $R = R_0$, and we have chosen the solutions so that Φ' is continuous at $R = R_0$. The value of Φ'_0 is fixed by the jump in $\mathrm{d}\Phi'/\mathrm{d}R$ across $R = R_0$. Integrating eq. (13.11) across $R = R_0$, we find that the jump is given by

$$\left[\frac{\mathrm{d}\Phi'}{\mathrm{d}R}\right]_{-}^{+} = 4\pi G\xi_R. \tag{13.17}$$

From this we obtain

$$\Phi'_0 = \frac{2\pi G\rho R_0}{m}\xi_R, \tag{13.18}$$

where the r.h.s. is to be evaluated at $R = R_0$.

We must now apply the free surface boundary condition at the surface of our 'star'. This requires $\delta p = 0$ at $R = R_0$, giving

$$\lim_{R \to R_0} \left(\frac{p'}{\rho} - \xi_R g\right) = 0, \tag{13.19}$$

where the surface gravity $g = \mathrm{d}\Phi/\mathrm{d}R$, evaluated at $R = R_0$, and $g > 0$. Alternatively we may write this as follows:

$$\lim_{R \to R_0} (W - \Phi' - \xi_R g) = 0. \tag{13.20}$$

To put this in a more useful form, we use eq. (13.18) to write Φ' in terms of ξ_R and then use the relationship $u_r = i\omega\xi_R$ together with eq. (13.13) to write ξ_R in terms of $\mathrm{d}W/\mathrm{d}R$. Note that in doing this we have used the equations which are valid for $R \neq R_0$, so that we have implicitly used the fact that $\lim_{R \to R_0} \rho' = 0$, even though the value of ρ' there is formally infinite! The boundary condition now becomes

$$\omega^2 W + \frac{dW}{dR}\left[-g + \frac{2\pi G\rho R_0}{m}\right] = 0, \tag{13.21}$$

all to be evaluated at $R = R_0$. Now from our solution for W given by eq. (13.15) we see that

$$\lim_{R \to R_0} \frac{1}{W}\frac{dW}{dR} = \frac{m}{R_0}. \tag{13.22}$$

Hence we find the mode frequencies as follows:

$$\omega^2 = \frac{mg}{R_0} - 2\pi G\rho, \tag{13.23}$$

provided $m \neq 0$. Of course the mode $m = 0$ represents a simple radial expansion or contraction of the cylinder, and cannot occur for an incompressible fluid. These modes correspond to surface gravity waves (g-modes).

13.1.1 Self-gravitating cylinder

If the cylinder is to be a proper model of a 'star', we must demand that the unperturbed gravitational potential Φ results solely from the gravity of the cylinder itself. Thus,

$$\frac{1}{R}\frac{d}{dR}\left(R\frac{d\Phi}{dR}\right) = 4\pi G\rho. \tag{13.24}$$

From this we can show that the surface gravity $g = d\Phi/dR$ is given by

$$g = 2\pi G\rho R_0. \tag{13.25}$$

Thus the oscillation frequencies of a self-gravitating cylinder are given by

$$\omega^2 = \frac{(m-1)g}{R_0}. \tag{13.26}$$

We note that although a mode with $m = 1$ is permitted, its frequency is zero. Since such a mode represents a simple translation of the star perpendicular to its axis of symmetry, this is exactly what we would expect. If you move a star bodily sideways, there is nothing to move it back.

13.2 Uniform rotation

We now consider the effect on the stellar modes of adding a small amount of uniform rotation Ω. There are two basic effects, which we discuss in Sections 13.2.1 and 13.2.2.

13.2.1 *Effect on p- and g-modes*

Rotation changes the equilibrium shape of the star by the addition of a centrifugal force into the equation of hydrostatic equilibrium. But this gives a term of order $\sim \Omega^2$. Thus to first order in the rotation rate Ω the basic structure of the star, and therefore its modes of oscillation, are not affected. However, there is a simple kinematic effect which is known as frequency splitting.

We found the azimuthal modes in the non-rotating cylinder to have the form

$$\xi_R \propto \sin(\omega t)\sin(m\phi), \tag{13.27}$$

or equivalently

$$\xi_R \propto \cos(\omega t - m\phi) - \cos(\omega t + m\phi). \tag{13.28}$$

We may view these as two surface waves, one moving in the positive ϕ-direction with phase velocity ω/m and the other moving in the negative ϕ-direction with phase velocity $-\omega/m$.

We have noted that the structure of a slowly rotating star is essentially unchanged. Therefore if we move with the rotating frame, the star is at rest and we can apply the analysis of the preceding section. The azimuthal coordinate ϕ' in the rotating frame is related to the azimuthal coordinate ϕ in the inertial frame by

$$\phi' = \phi - \Omega t. \tag{13.29}$$

We conclude therefore that for a slowly rotating star the modes take the form

$$\xi_R \propto \cos(\omega t - m\phi') - \cos(\omega t + m\phi'), \tag{13.30}$$

where the modal frequencies ω are unchanged to this order. Then in the inertial (non-rotating) frame the modes take the form

$$\xi_R \propto \cos[(\omega + m\Omega)t - m\phi] - \cos[(\omega - m\Omega)t + m\phi]. \tag{13.31}$$

Thus the effect of rotation is to split the frequency of the mode:

$$\omega \to \omega \pm m\Omega. \tag{13.32}$$

This is known as rotational splitting.

13.2.1.1 *The Chandrasekhar –Friedmann–Schutz instability*

The idea of rotational splitting leads to the remarkable conclusion that the emission of gravitational waves makes all rotating stars unstable! Einstein's general theory of relativity predicts that gravitational waves are emitted when a time-dependent density distribution gives rise to a time-dependent gravitational field. Gravitational waves have a quadrupole or higher multipole character, so that only modes with $m \geq 2$ produce them.

We consider the effect of gravitational wave emission on the modes we have been discussing.

First, we consider the non-rotating star. All the modes consist of two travelling waves, each moving with angular speed ω/m, one moving in the positive ϕ-direction and the other in the negative ϕ-direction (see eq. (13.28)). The one moving in the positive ϕ-direction has a positive associated angular momentum. The gravitational wave it emits also has positive angular momentum, so the effect of gravitational wave emission is to reduce the amplitude of the wave and thus damp it. Exactly the same argument applies to the wave moving in the negative ϕ-direction.

Now consider what happens if the star is rotating. The wave which was moving in the positive ϕ-direction is still moving in that direction, but now (in the inertial frame) with angular speed $\omega/m + \Omega$. Thus it still emits a gravitational wave with positive angular momentum, and therefore damps. But the wave which was moving in the negative ϕ-direction now has angular speed $-\omega/m + \Omega$ in that direction. Thus if $\omega < m\Omega$ this wave moves in the positive ϕ-direction in the inertial frame. If this happens, as it must for large enough m, then the wave, which has negative associated angular momentum, actually starts to emit gravitational radiation with *positive* angular momentum. This means that the wave amplitude grows and is the basis for what is called the Chandrasekhar–Friedmann–Schutz (CFS) instability.

In reality, damping processes occur fastest for the high-m modes, and this instability is not thought to operate for most stars. You can test this out by stamping on the ground (this should produce waves with $m \sim 10^9$) and seeing what happens! However, the instability is thought to play a role in limiting the rotation rate for neutron stars. These are very compact stars, with escape velocities from their surfaces around 10–20 per cent of the speed of light. This means that when they rotate rapidly, any surface features (such as waves) move at a speed v which is a significant fraction of the speed of light c and so become efficient emitters of gravitational waves (the emitted energy goes as $(v/c)^6$). Neutron stars can reach very rapid rotation rates by accreting matter with the Keplerian angular momentum at their equators. This gravitational wave instability would ultimately limit their spin rates if nothing else does at slower spin rates, and this is probably the reason that they appear to show a maximum rotation period of around 1.5 ms.

13.2.2 The r-modes

The second effect of introducing rotation is to give rise to a completely new set of modes, the rotational modes, or r-modes. Consider a perturbation of the form

$$\xi = (0, \xi_\phi(R, z), 0). \tag{13.33}$$

If the star is not rotating, then all this perturbation does is displace azimuthal rings of material in the ϕ-direction. Since the star is axi-symmetric, the equilibrium is unchanged and there is no restoring force which tries to put the fluid rings back where they were. In other words, the frequency associated with such a perturbation is $\omega = 0$. But if the star (or, in our case, cylinder) is rotating, then this displacement changes the angular velocity of the ring. This upsets the equilibrium of the star (which comes about through a balance between gravity, pressure gradient and centrifugal force) and so gives rise to oscillations. As we shall see, a typical frequency of the oscillations can be expected to be $\omega \sim \Omega$.

We return to eqs. (12.28) and (12.29) which we derived in Chapter 12. We assume all variables go as $\propto \exp\{i(\omega t + m\phi + kz)\}$. In Chapter 12 we defined the Rayleigh discriminant, which for the uniform rotation we consider here is given by

$$\mathcal{R} = 4\Omega^2, \tag{13.34}$$

and we defined the Doppler-shifted frequency,

$$\sigma = \omega + m\Omega. \tag{13.35}$$

We note that for uniform rotation, σ is independent of radius R. Then, in terms of the variables $W = p'/\rho$ and ξ_R, the linearized equations are given by

$$\frac{1}{R}\frac{d}{dR}(R\xi_R) - \frac{2m\omega}{\sigma R}\xi_R = \frac{1}{\sigma^2}\left(\frac{m^2}{R^2} + k^2\right)W \tag{13.36}$$

and

$$[\sigma^2 - 4\Omega^2]\xi_R = \frac{dW}{dR} + \frac{2m\Omega}{\sigma R}W. \tag{13.37}$$

We can now eliminate ξ_R between these two equations to get a second-order ordinary differential equation for W:

$$\frac{d^2W}{dR^2} + \frac{1}{R}\frac{dW}{dR} - \left[\left(1 - \frac{4\Omega^2}{\sigma^2}\right)k^2 + \frac{m^2}{R^2}\right]W = 0. \tag{13.38}$$

For a non-rotating incompressible cylinder, the only possible wave modes are surface waves (g-modes). To eliminate these we impose the boundary condition that $\xi_R = 0$ at the surface of the cylinder $R = R_0$. From eq. (13.37) we see that, in terms of W, this implies

$$\frac{dW}{dR} + \frac{2m\Omega}{\sigma R}W = 0 \tag{13.39}$$

at $R = R_0$.

We consider first what happens if the modes are independent of z, i.e. if $k = 0$. Without loss of generality, we may take $m \geq 0$. Then for $m \neq 0$ the solution which is finite at $R = 0$ is

$$W \propto R^m. \tag{13.40}$$

Since for this solution W and $\mathrm{d}W/\mathrm{d}R$ have the same sign, we cannot now satisfy the boundary condition, eq. (13.39). For $m = 0$ the solution that is finite at the origin is of the form $W = \text{const.}$ and corresponds to uniform radial expansion, which cannot occur for an incompressible fluid. We conclude that the r-modes must all have z-dependence.

We now need to recall briefly the basic properties of Bessel functions. This should not come as a surprise as they are the natural functions for describing oscillations in cylindrical geometry. The standard version of Bessel's equation for $y(x)$ is as follows:

$$\frac{\mathrm{d}^2 y}{\mathrm{d}x^2} + \frac{1}{x}\frac{\mathrm{d}y}{\mathrm{d}x} + \left(\alpha^2 - \frac{\nu^2}{x^2}\right) y = 0. \tag{13.41}$$

This has two independent solutions, which are oscillatory. These are $J_\nu(\alpha x)$, which is the cylindrical equivalent of $\sin x$ and (for $\nu \neq 0$) is zero at $x = 0$, and $Y_\nu(\alpha x)$, which is the cylindrical equivalent of $\cos x$ and is singular at $x = 0$. The modified version of Bessel's equation (obtained by the transformation $x \to ix$) is as follows:

$$\frac{\mathrm{d}^2 y}{\mathrm{d}x^2} + \frac{1}{x}\frac{\mathrm{d}y}{\mathrm{d}x} - \left(\alpha^2 + \frac{\nu^2}{x^2}\right) y = 0. \tag{13.42}$$

This has two independent solutions which are non-oscillatory – the modified Bessel functions. These are $I_\nu(\alpha x)$, which (for $\nu \neq 0$) is zero at $x = 0$ and is the cylindrical equivalent of $\exp x$, and $K_\nu(\alpha x)$, which is singular at $x = 0$ and is the cylindrical equivalent of $\exp(-x)$.

Returning to eq. (13.38), we see that if $\sigma^2 > 4\Omega^2$ then the solution which is finite at $x = 0$ has the form $W \propto I_m(\alpha R)$ for some real α. But in this case W and $\mathrm{d}W/\mathrm{d}R$ have the same sign, and we cannot then satisfy the boundary condition given in eq. (13.39). This tells us that the r-mode frequencies must satisfy $\sigma^2 < 4\Omega^2$, i.e.

$$-2\Omega < \sigma < 2\Omega. \tag{13.43}$$

When this is satisfied, we can define the real number α by

$$\alpha = \left[\frac{4\Omega^2}{\sigma^2} - 1\right]^{1/2} > 0. \tag{13.44}$$

Then the solution to eq. (13.38) is given by

$$W = J_m(\alpha k R). \tag{13.45}$$

Substituting this solution into the boundary condition, eq. (13.39), gives

$$\alpha k J'_m(\alpha k R_0) + \frac{2m\Omega}{\sigma R} J_m(\alpha k R_0) = 0, \tag{13.46}$$

where the prime denotes the derivative of the Bessel function with respect to its argument. Using the substitution

$$\frac{\sigma}{2\Omega} = \pm\frac{1}{\sqrt{1+\alpha^2}}, \tag{13.47}$$

we may rewrite this as follows:

$$\alpha k R_0 J_m'(\alpha k R_0) \pm m(1+\alpha^2)^{1/2} J_m(\alpha K R_0) = 0. \tag{13.48}$$

The oscillatory property of the Bessel function † means that for each value of k this gives multiple solutions for α and hence for σ or ω.

13.2.2.1 Axisymmetric r-modes

To make things simple, we consider the axisymmetric case $m = 0$. Then the boundary condition, eq. (13.39), becomes simply

$$J_0'(\alpha k R_0) = 0, \tag{13.49}$$

or equivalently, since from the properties of Bessel functions we know that $J_0'(x) \propto J_1(x)$,

$$J_1(\alpha k R_0) = 0. \tag{13.50}$$

This means that

$$\alpha k R_0 = \mathcal{Z}_i, \tag{13.51}$$

where $\mathcal{Z}_i$, for $i = 1, 2, 3, \ldots$, are the zeroes of the Bessel function $J_1(z)$. These can be looked up in tables, and the first few are $\mathcal{Z}_i = 3.83, 7.02, 10.17, \ldots$.

To make our cylinder look more like a star, we can assume that is has a finite height, so that it has fixed ends at $z = 0, H$. We then impose the further boundary conditions that $\xi_R = 0$ at $z = 0, H$, so that the z-dependence of W is $W \propto \sin(kz)$, where $kH = n\pi$ for non-zero integer n. After a little algebra we find that for such a 'star' the frequencies of the r-modes are given by

$$\omega_i = \pm\frac{2\Omega}{\left[1 + (\mathcal{Z}_i n\pi R_0/H)^2\right]^{1/2}}. \tag{13.52}$$

13.2.2.2 The CFS instability continued

Just as electromagnetic waves can be generated by time-varying electric currents, gravitational waves can be generated by mass currents. As we have seen, the r-modes consist essentially of oscillating mass flows within a rotating medium. It appears that it is these r-modes which are the most efficient drivers of the CFS instability in rapidly rotating neutron stars. One reason for this is that the modes do not require compressibility and do not damp as rapidly as compressible modes.

---

† If we imagine replacing J by sin and J' by cos, then this equation has the form $\tan A = B$ for some A and B, which of course gives multiple solutions for A.

13.3 Further reading

Rotational splitting of eigenfrequencies is discussed by Unno *et al.* (1979). The concept of rotational modes (r-modes) was introduced by Papaloizou & Pringle (1978a), who also introduced the concept of a minimum spin period for accreting neutron stars (Papaloizou & Pringle (1978b). They also note that since in three dimensions any disturbance can be described in terms of three independent vectors, the addition of the r-modes to the p- and g-modes implies that an arbitrary initial disturbance can be completely described in terms of these modes (see the discussion about waves in magnetic media, Chapter 2). The importance of singular modes of rotating stars (not discussed in this book, but see Section 10.3.1) in the stellar context is discussed by Ogilvie & Lin (2004) and Ogilvie (2005). The role of the continuous spectrum in rotational shear instabilities, and the use of the initial-value problem in determining the outcome, is discussed by Watts *et al.* (2003) and by Watts, Andersson & Williams (2004). A simple, pseudo-Newtonian description of the Chandrasekhar–Friedmann–Schutz (CFS) instability is given by Papaloizou & Pringle (1978b). The importance of r-modes in the CFS instability in the astrophysical context, specifically in rapidly rotating neutron stars, is presented by Andersson (1998).

13.4 Problems

13.4.1 A self-gravitating incompressible sphere has uniform density ρ, mass M and radius R. The star undergoes a perturbation which has zero vorticity and oscillation frequency ω. Show that this implies that we may write $\boldsymbol{\xi} = -\nabla \psi$ for some scalar field ψ which satisfies Laplace's equation $\nabla^2 \psi = 0$.

Show that the equation of motion then implies that

$$-\omega^2 \psi + \frac{p'}{\rho} + \Phi' = 0, \qquad (13.53)$$

where p' is the perturbation to the pressure and Φ' is the perturbation to the gravitational potential.

Deduce that p' and Φ' both satisfy Laplace's equation, so that we can write the solution in the following form:

$$\psi = C \left(\frac{r}{R}\right)^l Y_{lm}(\theta, \phi), \quad r < R,$$

$$p' = B \left(\frac{r}{R}\right)^l Y_{lm}(\theta, \phi), \quad r < R,$$

$$\Phi' = \begin{cases} A \left(\dfrac{r}{R}\right)^l Y_{lm}(\theta, \phi), & r < R, \\[2mm] A \left(\dfrac{R}{r}\right)^{l+1} Y_{lm}(\theta, \phi), & r > R, \end{cases} \qquad (13.54)$$

where A, B and C are constants and the $Y_{lm}(\theta, \phi)$ are spherical harmonics.

Use the equation of motion to find an equation relating A, B and C. Use the zero pressure boundary condition at $r = R$ to obtain a relation between B and C. Use Poisson's equation, applied as a jump condition at $r = R$, to relate A and C.

Hence show that the oscillation frequencies of a self-gravitating, uniform density, incompressible sphere are given by

$$\omega^2 = \frac{4}{3}\pi G\rho \frac{2l(l-1)}{2l+1}. \tag{13.55}$$

Why is $\omega = 0$ for $l = 0$ and for $l = 1$?

How does this result relate to the dispersion relation for deep water waves $\omega^2 = gk$?

13.4.2 (i) For the adiabatic oscillations of a spherical star, show that

$$\frac{d^2\boldsymbol{\xi}}{dt^2} = -\nabla\chi + \mathbf{A}\frac{\gamma p}{\rho}\,\text{div}\,\boldsymbol{\xi}, \tag{13.56}$$

where

$$\mathbf{A} = \frac{1}{\rho}\nabla p - \frac{1}{\gamma p}\nabla p \tag{13.57}$$

and

$$\chi = \frac{p'}{\rho} + \Phi'. \tag{13.58}$$

(ii) In spherical polar coordinates we write $\boldsymbol{\xi} = (\xi_r, \xi_\theta, \xi_\phi)$ and assume time-dependence of the form $e^{i\omega t}$. Show that

$$\text{div}\,\boldsymbol{\xi} = \frac{1}{r^2}\frac{\partial}{\partial r}(r^2\xi_r) - \frac{1}{\omega^2 r^2}L^2\chi, \tag{13.59}$$

where the operator L^2 is defined as follows:

$$L^2 = -\frac{1}{\sin\theta}\frac{\partial}{\partial\theta}\left(\sin\theta\frac{\partial}{\partial\theta}\right) - \frac{1}{\sin^2\theta}\frac{\partial^2}{\partial\phi^2}. \tag{13.60}$$

(iii) Let the perturbed quantities be of the form

$$\chi(r,\theta,\phi) = \chi(r)Y_{lm}(\theta,\phi). \tag{13.61}$$

Note that $\boldsymbol{\xi}$ is the sum of a purely radial vector and the gradient of a scalar. Show that if $\omega^2 \neq 0$, then $(\text{curl}\,\boldsymbol{\xi})_r = 0$. (These are the spheroidal modes, i.e. p-modes and g-modes.)

(iv) Give a physical explanation of the displacement

$$\boldsymbol{\xi} = e^{i\omega t}f(r)\left(0, \frac{1}{r\sin\theta}\frac{\partial Y_{lm}}{\partial\phi}, -\frac{1}{r}\frac{\partial Y_{lm}}{\partial\theta}\right), \tag{13.62}$$

where $f(r)$ is an arbitrary function.

Show that the displacement

$$\boldsymbol{\xi} = \hat{\mathbf{z}}\wedge\mathbf{r}\,e^{i\omega t} \tag{13.63}$$

is such a displacment for some $f(r)$, l and ω to be determined. (These are the toroidal modes, which become the r-modes for a rotating star.)

13.4.3 An unbounded incompressible fluid is rotating with constant angular velocity $\boldsymbol{\Omega}$ and is subject to a force $-\nabla\Phi$. Show that, in the rotating frame, a wave-like disturbance of the form $\exp\{i(\omega t + \mathbf{k} \cdot \mathbf{r})\}$ satisfies the equations

$$\mathbf{k} \cdot \mathbf{u} = 0 \tag{13.64}$$

and

$$i\omega\mathbf{u} + i(\mathbf{k} \cdot \mathbf{u})\mathbf{u} = -iB\mathbf{k} + 2\mathbf{u} \wedge \boldsymbol{\Omega}, \tag{13.65}$$

where the scalar B and the vector $\mathbf{u}$ are both complex quantities.

Note that $\mathbf{u} \cdot \mathbf{u} = 0$ does not imply that $\mathbf{u} = 0$.

Assuming that we are considering propagating waves, so that $\mathbf{k}$ and ω are real, show that

$$\omega = \pm 2\boldsymbol{\Omega} \cdot \hat{\mathbf{k}}. \tag{13.66}$$

Deduce that the group velocity is given by

$$\frac{\partial\omega}{\partial\mathbf{k}} = \pm\frac{2}{k^3}\mathbf{k} \wedge (\boldsymbol{\Omega} \wedge \mathbf{k}). \tag{13.67}$$

Describe the properties of the waves.

Compare the properties of these waves with the properties of buoyancy waves in an incompressible stably stratified fluid (Chapter 5). (See Greenspan (1968, Chap. 4).)

13.4.4 In cylindrical polar coodinates (R, ϕ, z) an incompressible fluid of uniform density is in circular flow with angular velocity

$$\Omega(R) = \begin{cases} \Omega_0, & R < R_0, \\ \Omega_0(R/R_0)^{-2}, & R > R_0, \end{cases} \tag{13.68}$$

where R_0 and Ω_0 are constants. The flow is now subject to small perturbations of the form $f(R)\exp\{i(\omega t + m\phi)\}$, where $m \neq 0$. In the domain $0 \leq R \leq R_0$, show that $W = p'/\rho$ has solutions $W \propto R^m$. In the domain $R_0 \leq R$, show that $W \propto (\omega + m\Omega)R^{-m}$.

Hence show that the oscillation frequencies are given by $\omega = \Omega_0(1 - m)$. (See Lamb (1932, Chap. VII).)

13.4.5 An incompressible fluid of uniform density ρ rotates uniformly with angular velocity Ω inside a smooth rigid sphere of radius r_0. Find the pressure distribution in the fluid.

The flow undergoes small oscillations such that, for example, the Eulerian pressure perturbation is of the form

$$p' = p'(R, z)\exp\{i(\omega t + m\phi)\}, \tag{13.69}$$

where (R, ϕ, z) are cylindrical polar coordinates, m is the azimuthal wavenumber and ω is the oscillation frequency. Show that $W = p'/\rho$ satisfies the equation

$$\frac{1}{R}\frac{\partial}{\partial R}\left(R\frac{\partial W}{\partial R}\right) - \frac{m^2}{R^2}W + \left(1 - \frac{4\Omega^2}{\sigma^2}\right)\frac{\partial^2 W}{\partial z^2} = 0, \tag{13.70}$$

where $\sigma = \omega + m\Omega$.

Derive the boundary condition satisfied by W on the sphere $R^2 + z^2 = r_0^2$.
Show that there is an axisymmetric oscillation mode of the form

$$W = z(AR^2 + Bz^2 + C), \tag{13.71}$$

where A, B and C are constants, with oscillation frequency given by

$$\omega = \pm \frac{\sqrt{5}}{2}\Omega. \tag{13.72}$$

14

Cylindrical shear flow–non-axisymmetric instability

In several astrophysical contexts, notably in the study of accretion discs, we would like to know whether certain types of rotating shear flow are stable. If so, these are potential candidates for realistic models of accretion flows. In Chapter 12 we considered cylindrical shear flow in an incompressible fluid. We showed that in the absence of magnetic fields the flow is unstable to axisymmetric instability if the specific angular momentum is a decreasing function of radius – the Rayleigh criterion. Thus we need to ask if the converse is true; that is, whether stability is guaranteed if the Rayleigh criterion is satisfied.

We shall show here, by considering a particular example, that this converse is not true. If non-axisymmetric perturbations are allowed, instability is still possible even if the Rayleigh criterion is satisfied.

14.1 Equilibrium configuration

We consider a cylindrical flow of an incompressible fluid with no z-dependence in a fixed gravitational potential:

$$\Phi(R) = -\frac{GM}{R}. \tag{14.1}$$

Note that this is a purely artificial potential, since R is the cylindrical radius not the spherical radius r which appears in the gravitational potential $\Phi = -GM/r$ for a point mass M at the origin. Since there is no z-dependence, the equilibrium equation is simply

$$-\frac{1}{\rho}\nabla p - \nabla\Phi + R\Omega^2\hat{\mathbf{R}} = 0. \tag{14.2}$$

Here $\Omega(R)$ is the angular velocity of the fluid. Replacing the centrifugal force term by a rotational potential of the form $\Phi_\Omega(R)$, where

$$\frac{\partial\Phi_\Omega}{\partial R} = -R\Omega^2, \tag{14.3}$$

191

we can integrate the equilibrium equation to give

$$\frac{p}{\rho} + \Phi + \Phi_\Omega = C, \tag{14.4}$$

where C is a constant.

We now specialize the flow to one for which the Rayleigh criterion is neutrally satisfied, i.e. one for which $\mathcal{R}(R) = 0$. Thus we have

$$\Omega(R) = \Omega_0 \left(\frac{R_0}{R}\right)^2, \tag{14.5}$$

where Ω_0 is the angular velocity at some reference radius R_0, yet to be defined. We note that in eq. (14.2) the gravitational force has the form $F_G \propto -1/R^2$ and the centrifugal force obeys $F_\Omega \propto 1/R^3$. This means that at small radii the centrifugal force dominates and the net force is outwards, while at large radii the gravitational force dominates and the net force is inwards. Therefore at some radius, which we define as the reference radius R_0, these two balance and we have $F_G + F_\Omega = 0$. Using this definition we find that

$$R_0 \Omega_0^2 = \frac{GM}{R_0^2}, \tag{14.6}$$

and hence that we can rewrite eq. (14.4) in the following form:

$$\frac{p(R)}{\rho} = C + \frac{GM}{R} - \frac{GMR_0}{2R^2}. \tag{14.7}$$

For a given value of the constant C, this defines the distribution of pressure $p(R)$. We note that by definition we have ensured that $\mathrm{d}p/\mathrm{d}R = 0$ at $R = R_0$, and indeed $p(R)$ reaches a maximum value of

$$p_{\mathrm{max}} = \rho \left(C + \frac{GM}{2R_0}\right) \tag{14.8}$$

at that radius. For $R > R_0$, $p(R)$ is a monotonically decreasing function of R, reaching the value of $p_\infty = \rho C$ in the limit as $R \to \infty$. From eq. (14.7) we see that, as R decreases, $p(R)$ always becomes negative at some finite radius. Since in the body of the fluid the pressure must always be positive, this implies that the fluid must have a cylindrical cavity surrounding the $z = 0$ axis.

We are interested here in finite configurations, which do not reach to infinity. This means that we require $C < 0$, so that there is no pressure at infinity. Also, for the configuration to exist at all, we require that $p_{\mathrm{max}} > 0$. We shall therefore redefine C in terms of a dimensionless parameter λ as follows:

$$C = -\frac{GM}{\lambda R_0}, \tag{14.9}$$

where, for a finite fluid configuration, we now require $2 < \lambda < \infty$. To locate the boundaries of the fluid configuration, we look for the radii at which $p(R) = 0$. Substituting all this into eq. (14.7) we find that the fluid configuration is in the region $R_- \leq R \leq R_+$, where $R_\pm$ are the roots of a quadratic equation given by

$$\frac{R_\pm}{R_0} = \frac{1}{2}(\lambda \pm \sqrt{\lambda(\lambda - 2)}). \tag{14.10}$$

We note that we may define $R_- < R_0$ and $R_+ > R_0$. As $\lambda \to 2$, the two roots become almost equal and the fluid configuration is a slender cylindrical shell. As $\lambda \to \infty$, we see that $R_- \to 1/2$ while $R_+ \to \infty$.

Finally, for later use, we note that the effective gravity in the flow, $g = \nabla p/\rho$, can be written as follows:

$$g(R) = \frac{GM}{R_0^2} \left[\frac{R_0}{R}\right]^3 \left[1 - \frac{R}{R_0}\right]. \tag{14.11}$$

Then of course at the inner radius, $R = R_-$, the gravity $g(R_-) = g_- > 0$ acts radially outwards, while at the outer radius, $R = R_+$, the gravity $g(R_+) = g_+ < 0$ acts radially inwards.

14.2 The perturbation equations

We proceed exactly as before to obtain the linearized equations of motion, noting that in this case we assume that there is no z-dependence. We also ignore the self-gravity of the fluid, so we assume the potential to be fixed, and note that the Eulerian density perturbation ρ' is zero everywhere except for a δ-function contribution at the boundaries. As before, we take all quantities to vary as $\propto \exp\{i(\omega t + m\phi)\}$ and define the Doppler-shifted frequency

$$\sigma(R) = \omega + m\Omega(R). \tag{14.12}$$

Then in terms of the radial Lagrangian displacement ξ_R and the quantity $W = p'/\rho$, and recalling that here $\Omega \propto 1/R^2$, eqs. (12.28) and (12.29) become

$$\sigma^2 \left(\frac{d\xi_R}{dR} + \frac{\xi_R}{R}\right) - \frac{2m\Omega\sigma}{R}\xi_R = \frac{m^2}{R^2}W \tag{14.13}$$

and

$$\sigma^2 \xi_R = \frac{dW}{dR} + \frac{2m\Omega}{\sigma R}W. \tag{14.14}$$

Because $\Omega \propto 1/R^2$, these equations can be simplified by using the substitution

$$W = \sigma H. \tag{14.15}$$

Then eq. (14.14) becomes simply

$$\sigma \xi_R = \frac{dH}{dR},\tag{14.16}$$

and eq. (14.13) becomes

$$\frac{d}{dR}(\sigma \xi_R) + \sigma \frac{\xi_R}{R} = \frac{m^2}{R^2}H.\tag{14.17}$$

Eliminating ξ_R from these two equations, we obtain

$$\frac{1}{R}\frac{d}{dR}\left(R\frac{dH}{dR}\right) - \frac{m^2}{R^2}H = 0.\tag{14.18}$$

This is a second-order equation for H with eigenvalues ω, once we have applied appropriate boundary conditions. We know from Chapter 4 that if we have fixed boundaries $\xi_R = 0$ there are no surface waves (the incompressible equivalent of p- and g-modes). In addition, since we have no z-dependence, then there are no internal waves (r-modes). Thus, in order to allow something to happen we must let the boundaries move. The simplest boundary condition, and the most appropriate one from an astrophysical point of view, is to allow free boundaries at $R = R_\pm$. Thus at $R = R_\pm$ we let

$$\delta p = p' + \xi_R \frac{dp}{dR} = 0.\tag{14.19}$$

This implies that

$$W + g\xi_R = 0,\tag{14.20}$$

which, in turn, using the definition of H and eq. (14.16), gives

$$\sigma^2 H + g\frac{dH}{dR} = 0,\tag{14.21}$$

which is valid at $R = R_\pm$.

The general solution to eq. (14.18) is as follows:

$$H = C_1 R^m + C_2 R^{-m},\tag{14.22}$$

where C_1 and C_2 are constants. Applying the boundary condition, eq. (14.21), at $R = R_-$ and $R = R_+$, and then eliminating the constants C_1 and C_2, we obtain the eigenvalue equation for ω:

$$\frac{(\omega + m\Omega_-)^2 + mg_-/R_-}{(\omega + m\Omega_+)^2 + mg_+/R_+} = \left(\frac{R_+}{R_-}\right)^{2m}\frac{(\omega + m\Omega_-)^2 - mg_-/R_-}{(\omega + m\Omega_+)^2 - mg_+/R_+},\tag{14.23}$$

where we have written $\Omega_\pm = \Omega(R_\pm)$. This is a quartic equation for ω.

14.3 The Papaloizou–Pringle instability

The roots of the quartic equation,eq. (14.23), are able to give rise to unstable growing modes, i.e. values of ω such that $\text{Im}(\omega) < 0$. This happens for the following reason. At the inner edge there are two sets of waves, travelling relative to the fluid in the positive and negative ϕ-directions. The same applies at the outer edge. However, because $\Omega \propto 1/R^2$, the fluid at the inner edge is moving at an angular velocity Ω_- greater than the angular velocity Ω_+ at the outer edge. The instability occurs when a wave which is travelling backwards relative to the fluid at the inner edge has the same angular phase speed as a wave which is travelling forwards relative to the fluid at the outer edge. When this happens the waves can communicate and exchange angular momentum and energy . The situation here is exactly analogous to the one we discussed in Chapter 13 regarding the Chandrasekhar –Friedmann– Schutz instability. The wave at the inner edge transfers angular momentum outwards to the wave at the outer edge. The wave at the outer edge has positive angular momentum and is *gaining* angular momentum so it grows in amplitude. The wave at the inner edge which is losing angular momentum is a wave of negative angular momentum, and so it too grows in amplitude. This signifies an instability, here called the Papaloizou–Pringle (PP) instability.

Here we consider two limiting cases, amenable to simple analysis, one stable and the other unstable.

14.3.1 Large m

Here we consider what happens in the limit of large m. In this limit the wavelengths of the surface waves ($2\pi R_{\pm}/m$ in the azimuthal direction) become very short. Thus they behave like deep-water waves and have frequencies (in the frame of the fluid) given by $\omega^2 = mg_{\pm}/R_{\pm}$. In eq. (14.23), the term $(R_+/R_-)^{2m}$ becomes very large in the limit of large m and dominates all the other terms. For equality to hold therefore, we require either that the denominator of the l.h.s. or the numerator of the r.h.s. vanish. These two conditions, recalling that $g_+ < 0$ and $g_- > 0$, give the four roots of the quadratic:

$$\omega = -m\Omega_+ \pm \sqrt{m|g_+|/R_+}, \tag{14.24}$$

which correspond to the expected surface waves on the outer edge, and

$$\omega = -m\Omega_- \pm \sqrt{mg_-/R_-}, \tag{14.25}$$

which correspond to surface waves on the inner edge. For such surface waves, the phase velocities are $\propto 1/\sqrt{m}$ so there is no chance of the backward wave at the inner surface being in phase with the forward wave on the outer surface. Thus in this limit the waves are stable. Moreover at such large values of m, the depth to which such waves propagate means that the waves on the two surfaces do not interact and move independently.

14.3.2 Thin cylindrical shell

We consider the limit $\lambda \to 2$, so that the fluid is confined to a thin cylindrical shell. We do this at a fixed value of m and this now gives the waves a chance to interact. Thus we set

$$\lambda = 2(1 + \epsilon^2) \tag{14.26}$$

and consider the limit $\epsilon \ll 1$. To first order in ϵ, we find that

$$R_\pm = R_0(1 \pm \epsilon), \tag{14.27}$$

that

$$g_\pm = \pm \epsilon R_0 \Omega_0^2 \tag{14.28}$$

and that

$$\Omega_\pm = \Omega_0(1 \mp 2\epsilon). \tag{14.29}$$

It is convenient to define

$$\sigma_0 = \omega + m\Omega_0, \tag{14.30}$$

so that

$$\omega + m\Omega_\pm = \omega + m\Omega_0 \mp 2m\Omega_0\epsilon. \tag{14.31}$$

Substituting this into eq. (14.23) and keeping only first-order terms in ϵ, the equation reduces to

$$\sigma^2 = -2\Omega_0^2. \tag{14.32}$$

This implies

$$\omega = -m\Omega_0 \pm i\Omega_0\sqrt{2}. \tag{14.33}$$

We note two things about this result. First, the real part of the eigenfrequency is such that the eigenmode co-rotates with the fluid at radius $R = R_0$. It is essential that unstable modes co-rotate with the fluid at some radius, so that forward and backward travelling waves have a chance to interact. We already found that this was a requirement for instability in the linear shear flow (Chapter 10). Second, the growth rate of the instability is dynamical, with the growth timescale being of order the rotation period of the fluid. Thus, even though according to the Rayleigh criterion the fluid is on the margin of stability to axisymmetric modes, it is actually dynamically unstable to non-axisymmetric modes. This demonstrates the distinction between local and global instability. The Rayleigh criterion is purely local, whereas the PP instability explicitly involves the conditions at the inner and outer boundaries of the flow. In astrophysical applications the PP instability shows that the possible forms of rotating shear flows in accretion discs are severely restricted.

14.4 Further reading

The original description of the Papaloizou–Pringle instability is to be found in Papaloizou & Pringle (1984). The simplified analysis presented here is based on Blaes & Glatzel (1986).

14.5 Problems

14.5.1 A torus of non-self-gravitating polytropic fluid (index n) rotates about the $R = 0$ axis with uniform specific angular momentum h and is subject to the gravitational field of a point mass M at the origin $R = z = 0$. Show that the structure of the torus is given by

$$(n+1)\frac{p}{\rho} = \frac{GM}{R_0}\left[\left(\frac{R_0^2}{R^2 + z^2}\right)^{1/2} - \frac{1}{2}\left(\frac{R_0}{R}\right)^2 - C\right], \qquad (14.34)$$

where C is a constant and the density maximum is at $R = R_0$, $z = 0$, where $h^2 = GMR_0$.

Describe approximately the shapes of the tori for values of C in the range $0 < C < 1/2$. If $C \approx 1/2$ show that the torus is very slender and has a nearly circular cross section. (See Papaloizou & Pringle (1984).)

14.5.2 Consider the torus discussed in Problem 14.5.1. Show that small perturbations of the form $\propto \exp\{i(\omega t + m\phi)\}$ obey the equation

$$\frac{1}{R}\frac{\partial}{\partial R}\left(\rho R\frac{\partial W}{\partial R}\right) = -\frac{m^2}{R^2}\rho W + \frac{\partial}{\partial z}\left(\rho\frac{\partial W}{\partial z}\right) = -\frac{\sigma^2\rho^2}{\gamma p}W, \qquad (14.35)$$

where $W = p'/\rho\sigma$ and $\sigma = \omega + m\Omega$.

14.5.3 An infinite cylinder ($0 \le R \le R_0$) of incompressible fluid with uniform density ρ_0 rotates about the axis $R = 0$ with velocity $\mathbf{u}_0 = (0, R\Omega(R), 0)$, with $\Omega(R) = kR$, where k is a constant.

The fluid is self-gravitating. Show that if the central pressure $p(R{=}0) = \pi^2 G^2\rho_0^3/k^2$, then the radius is $R_0 = (2\pi G\rho_0)^{1/2}/k$ and the effective surface gravity is zero.

The fluid is subject to small perturbations so that, for example, the velocity is $\mathbf{u}_0 + \mathbf{u}$, where $\mathbf{u}$ is of the form

$$\mathbf{u} \propto (u_R(R), u_\phi(R), 0)\exp\{i(\omega t + m\phi)\}. \qquad (14.36)$$

Show that the perturbation equations are given by

$$i\sigma u_R - 2\Omega u_\phi = -\frac{dW}{dR}, \qquad (14.37)$$

$$3\Omega u_R + i\sigma u_\phi = -\frac{im}{R}W \qquad (14.38)$$

and

$$\frac{du_R}{dR} + \frac{u_R}{R} + \frac{im}{R}u_\phi = 0, \tag{14.39}$$

where $\sigma = \omega + m\Omega(R)$ and $W = p'/\rho + \Phi'$.

Show that these equations can be reduced to the following single equation:

$$\frac{d^2 u_R}{dR^2} + \frac{3}{R}\frac{du_R}{dR} + \left\{1 - m^2 - \frac{3m\Omega}{\sigma}\right\}\frac{u_R}{R^2} = 0. \tag{14.40}$$

Now consider the case $m = 1$. Show that a solution to this equation is given by

$$u_R = 1 + \frac{kR}{\omega}. \tag{14.41}$$

Assuming that this is the only solution which is regular at $R = 0$, show that the oscillation frequencies obey the equation $\omega^2 = 0$. Give a physical explanation of this result.

In this case the solution given is apparently singular. Find the solution to eq. (14.40) for the case $m = 1$ and $\omega = 0$, which is regular at the origin, and comment on how it is obtained from the solution given in the limit $\omega \to 0$. (See Watts *et al.* (2003, 2004), and also refer to Balbinski (1984).)

References

Andersson, N. (1998). A new class of unstable modes of rotating relativistic stars, *Astrophys. J.*, **502**, 708.

Balbinski, E. (1984). The continuous spectrum in differentially rotating perfect fluids: a model with an analytic solution, *Mon. Not. Royal Astron. Soc.*, **209**, 145.

Balbus, S. A. & Hawley, J. F. (1998). Instability, turbulence, and enhanced transport in accretion disks, *Rev. Mod. Phys.*, **70**, 1.

Batchelor, G. K. (1967). *An Introduction to Fluid Dynamics* (Cambridge: Cambridge University Press).

Billingham, J. & King, A. C. (2000). *Wave Motion* (Cambridge: Cambridge University Press).

Blaes, O. M. & Glatzel, W. (1986). On the stability of incompressible constant angular momentum cylinders, *Mon. Not. Royal Astron. Soc.*, **220**, 253.

Blandford, R. D. & Rees, M. J. (1974). A 'twin-exhaust' model for double radio sources, *Mon. Not. Royal Astron. Soc.*, **169**, 395.

Bondi, H. (1952). On spherically symmetrical accretion, *Mon. Not. Royal Astron. Soc.*, **112**, 195.

Chandrasekhar, S. (1961). *Hydrodynamic and Hydromagnetic Stability* (Oxford: Oxford University Press); republished in 1981 by Dover Publications, Inc., New York.

Clayton, D. D. (1983). *Principles of Stellar Evolution and Nucleosynthesis* (Chicago: University of Chicago Press).

Coles, P. & Lucchin. (1995). *Cosmology – The Origin and Evolution of Cosmic Structure* (New York: John Wiley & Sons).

Cox, J. P. (1980). *Theory of Stellar Pulsation* (Princeton: Princeton University Press).

Doroshkevich, A. G. (1980). Fragmentation of a primordial flat layer, and the formation of internal cluster structure, *Soviet Astronomy*, **24**, 152.

Drazin, P. & Reid, W. (1981). *Hydrodynamic Stability* (Cambridge: Cambridge University Press).

Elphick, C., Regev, O. & Shaviv, N. (1992). Dynamics of fronts in thermally bistable fluids, *Astrophys. J.*, **392**, 106.

Elphick, C., Regev, O. & Spiegel, E. A. (1991). Complexity from thermal instability, *Mon. Not. Royal Astron. Soc.*, **250**, 617.

Field, G. B. (1965). Thermal instability, *Astrophys. J.*, **142**, 531.

Field, G. B., Rather, J. D. G., Aannenstad, P. A. & Orszag, S. A. (1968). Hydromagnetic shock waves and their infrared emission in HI regions, *Astrophys. J.*, **151**, 953.

Garaud, P. (2002). On rotationally driven meridional flows in stars, *Mon. Not. Royal Astron. Soc.*, **335**, 707.

Gerwin, R. A. (1968). Stability of the interface between two fluids in relative motion, *Rev. Mod. Phys.*, **40**, 652.

Greenspan, H. P. (1968). *The Theory of Rotating Fluids* (Cambridge: Cambridge University Press).

Howard, L. N. (1961). Note on a paper of John W. Miles, *J. Fluid Mech.*, **10**, 509.

Jackson, J. D. (1998). *Classical Electrodynamics*, 3rd edn (New York: John Wiley & Sons, Inc.).

Jeans, Sir J. H. (1929). *Astronomy and Cosmogony*, 2nd edn (Cambridge: Cambridge University Press); republished in 1961 by Dover Publications, Inc., New York.

Königl, A. (1982). On the nature of bipolar sources in dense molecular clouds, *Astrophys. J.*, **261**, 115.

Lamb, Sir H. (1932). Hydrodynamics, 6th edn (Cambridge: Cambridge University Press); republished in 1945 by Dover Publications, Inc., New York.

Landau, L. D. & Lifshitz, E. M. (1959). *Fluid Mechanics* (Oxford: Pergamon Press).

Ledoux, P. (1951). Sur la stabilité gravitationelle d'une nébuleuse isotherme, *Ann. Astrophys.*, **14**, 438.

Lifshitz, E. M. & Pitaevskii, L. P. (1980). *Statistical Physics*, 3rd edn, part 1 (Oxford: Pergamon Press).

Lubow, S. H. & Pringle, J. E. (1993). The gravitational stability of a compressed slab of gas, *Mon. Not. Royal Astron. Soc.*, **263**, 701.

Meerson, B. (1996). Non-linear dynamics of radiative condensations in optically thin plasmas, *Rev. Mod. Phys.*, **68**, 215.

Miles, J. W. (1961). On the stability of heterogeneous shear flows, *J. Fluid Mech.*, **10**, 496.

Ogilvie, G. I. & Lin, D. N. C. (2004). Tidal dissipation in rotating giant planets, *Astrophys. J.*, **610**, 477.

Ogilvie, G. I. (2005). Wave attractors and the asymptotic dissipation rate of tidal disturbances, *J. Fluid Mech.*, **543**, 16.

Papaloizou, J. & Pringle, J. E. (1978a). Non-radial oscillations of rotating stars and their relevance to the short period oscillations of cataclysmic variables, *Mon. Not. Royal Astron. Soc.*, **182**, 423.

Papaloizou, J. & Pringle, J. E. (1978b). Gravitational radiation and the stability of rotating stars, *Mon. Not. Royal Astron. Soc.*, **184**, 501.

Papaloizou, J. & Pringle, J. E. (1984). The dynamical stability of differentially rotating discs with constant specific angular momentum, *Mon. Not. Royal Astron. Soc.*, **208**, 721.

Parker, E. N. (1979). *Cosmical Magnetic Fields – Their Origin and Their Activity* (Oxford: Clarendon Press).

Pringle, J. E. & Rees, M. J. (1972). Accretion disc models for compact x-ray sources, *Astron. Astrophys.*, **21**, 1.

Roberts, P. H. (1967). *An Introduction to Magnetohydrodynamics* (London: Longmans, Green and Co. Ltd).

Sedov, L. (1959). *Similarity and Dimensional Methods in Mechanics* (New York: Academic Press).

Shu, F. H. (1992). *Gas Dynamics* (Millvalley, CA: University Science Books).

Sturrock, P. A. (1994). *Plasma Physics – An Introduction to the Theory of Astrophysical, Geophysical and Laboratory Plasmas* (Cambridge: Cambridge University Press).

Tassoul, J.-L. (1978). *Theory of Rotating Stars* (Princeton: Princeton University Press).

Tassoul, J.-L. (2000). *Stellar Rotation* (Cambridge: Cambridge University Press).

Taylor, G. I., (1959a). The formation of a blast wave by a very intense explosion. II. The atomic explosion of 1945, *Proc. Roy. Soc. London A*, **201**, 159.

Taylor, G. I., (1959b). The formation of a blast wave by a very intense explosion. I. Theoretical discussion, *Proc. Roy. Soc. London A*, **201**, 175.

Toomre, A. (1964). On the gravitational stability of a disk of stars, *Astrophys. J.*, **139**, (1217).

Turner, J. S. (1973). *Buoyancy Effects in Fluids* (Cambridge: Cambridge University Press).

Unno, W., Osaki, Y., Ando, H. & Shibahashi, H. (1979). *Non-Radial Oscillations of Stars* (Tokyo: University of Tokyo Press).

Watts, A. L., Andersson, N. & Williams, R. L. (2004). The oscillation and stability of differentially rotating spherical shells: the initial-value problem, *Mon. Not. Royal Astron. Soc.*, **350**, 927.

Watts, A. L., Andersson, N., Beyer, H. & Schutz, B. F. (2003). The oscillation and stability of differentially rotating spherical shells: the normal-mode problem, *Mon. Not. Royal Astron. Soc.*, **342**, 1156.

Weinberg, S. (1972). *Gravitation and Cosmology – Principles and Applications of the General Theory of Relativity* (New York: John Wiley & Sons, Inc.).

Witham, G. B. (1974). *Linear and Non-Linear Waves* (New York: John Wiley & Sons).

Zel'dovich, Ya. B. & Novikov, I. D. (1971). In K. S. Thorne & W. D. Arnett, *Relativistic Astrophysics, Vol. 1: Stars and Relativity* (Chicago: University of Chicago Press).

Zel'dovich, Ya. B. & Raizer, Yu. P. (1967). *Physics of Shockwaves and High-Temperature Hydrodynamic Phenomena* (New York: Academic Press).

Index

Printed in the United States
By Bookmasters